Introduction to Cell Biology

Introduction to Cell Biology

Edited by

Casper Barnes

Larsen & Keller
www.larsen-keller.com

Introduction to Cell Biology
Edited by Casper Barnes
ISBN: 978-1-63549-061-9 (Hardback)

Larsen & Keller
Published by Larsen and Keller Education,
5 Penn Plaza,
19th Floor,
New York, NY 10001, USA

Cataloging-in-Publication Data

Introduction to cell biology / edited by Casper Barnes.
p. cm.
Includes bibliographical references and index.
ISBN 978-1-63549-061-9
1. Cytology. 2. Cells. I. Barnes, Casper.
QH581.2 .I68 2017
571.6--dc23

Printed and bound in China.

For more information regarding Larsen and Keller Education and its products, please visit the publisher's website www.larsen-keller.com

Table of Contents

Preface

Cell biology is that branch of biology which studies the organization, structure, physiological properties, life cycle, metabolic processes, signaling pathways of cells and their interaction with the environment. It overlaps with developmental biology, immunology, biochemistry, etc. This book attempts to understand the multiple branches that fall under the discipline of cell biology and how such concepts have practical applications. Such selected concepts that redefine the subject have been presented in the book. For all those who are interested in the subject, this text can prove to be an essential guide.

To facilitate a deeper understanding of the contents of this book a short introduction of every chapter is written below:

Chapter 1- Cell biology discusses all the essential parts as well as functions that make up a cell. A cell is the most basic unit of life. The chapter on cell biology offers an insightful focus, keeping in mind the complex subject matter.

Chapter 2- Cells carry out different functions and are arranged according to the role that they play in an organism's lifetime. Cells can be classified as single-celled or multi-celled correlated to the evolutionary stage of the life form. Cell function also changes according to these differences.

Chapter 3- An evolved cell has a particular structure that houses components such as the cell nucleus, golgi apparatus and the cytoskeleton. All these contribute to the functioning of the cell. The topics discussed in this chapter are of great importance to broaden the existing knowledge on cell structure.

Chapter 4- Our understanding of cells and cell function come from theories that were promulgated by scholars through the years. In order to study cell and cell structure in detail, it is vital to know the basic stages of the cell cycle as well as theories on its origin. This chapter strategically encompasses and incorporates the major components and key concepts of cell biology, providing a complete understanding.

Chapter 5- Cells originate and reproduce through cell division. All cells reach a stage where the cell divides into daughter cells. This is an important area of study in cell biology. This chapter will provide an integrated understanding of cell reproduction.

Chapter 6- Cell function and cell metabolism require certain fundamental activities to survive such as the ingestion and digestion of food and the flow of blood, in the case of animals. Cells also perform complex actions that can be defined under cell adhesion.

Chapter 7- There are many theories regarding the origin and evolution of cells. One of the popular ones is symbiogenesis, which regards the two main cell types, prokaryotes and eukaryotes as interacting with each other and leading to the differentiation and classification of cells into the two groups. This chapter elucidates the crucial theories and principles of symbiogenesis.

Chapter 8- Cell biology is very important in the fields of genetics and pharmaceutical sciences. Experiments on drug effectiveness and their side-effects as well as newer techniques of cloning and gene splicing all began with the study of cell biology. The topics discussed in the chapter are of great importance to broaden the existing knowledge on cell biology.

Finally, I would like to thank the entire team involved since the inception of this book for their valuable time and contribution. This book would not have been possible without their efforts. I would also like to thank my friends and family for their constant support.

Editor

1

Introduction to Cell Biology

Cell biology discusses all the essential parts as well as functions that make up a cell. A cell is the most basic unit of life. The chapter on cell biology offers an insightful focus, keeping in mind the complex subject matter.

Cell biology is a branch of biology that studies the different structures and functions of the cell and focuses mainly on the idea of the cell as the basic unit of life. Cell biology explains the structure, organization of the organelles they contain, their physiological properties, metabolic processes, signaling pathways, life cycle, and interactions with their environment. This is done both on a microscopic and molecular level as it encompasses prokaryotic cells and eukaryotic cells. Knowing the components of cells and how cells work is fundamental to all biological sciences; it is also essential for research in bio-medical fields such as cancer, and other diseases. Research in cell biology is closely related to genetics, biochemistry, molecular biology, immunology, and developmental biology.

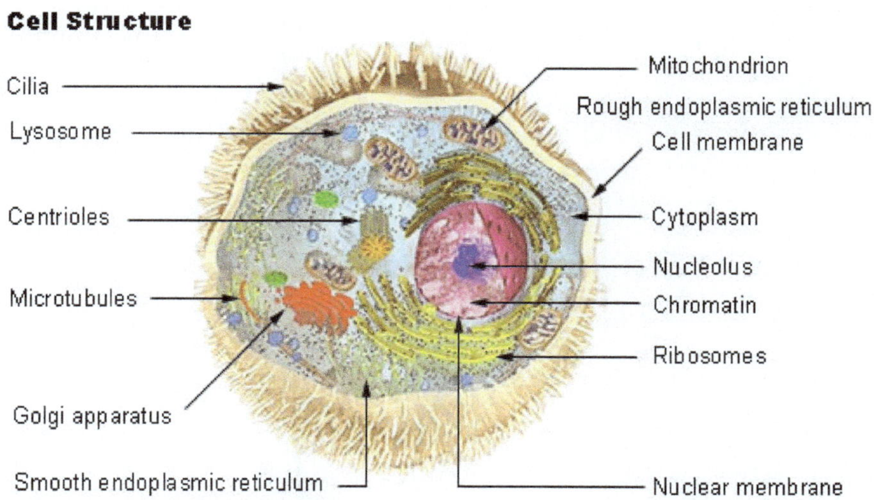

Understanding the cell in terms of its generalized structure and molecular components.

Internal Cellular Structures

Chemical and Molecular Environment

The study of the cell is done on a molecular level; however, most of the processes within the cell are made up of a mixture of small organic molecules, inorganic ions, hormones, and water. Approximately 75-85% of the cell's volume is due to water making it an indispensable solvent as a result of its polarity and structure. These molecules within the cell, which operate as substrates, provide a suitable environment for the cell to carry out metabolic reactions and signalling. The cell shape varies among the different types of organisms, and are thus then classified into two categories: eukaryotes and prokaryotes. In the case of eukaryotic cells - which are made up of animal, plant, fungi, and protozoa cells - the shapes are generally round and spherical, while for prokaryotic cells – which are composed of bacteria and archaea - the shapes are: spherical (cocci), rods (bacillus), curved (vibrio), and spirals (*spirochetes).*

Cell biology focuses more on the study of eukaryotic cells, and their signalling pathways, rather than on prokaryotes which is covered under microbiology. The main constituents of the general molecular composition of the cell includes: proteins and lipids which are either free flowing or membrane bound, along with different internal compartments known as organelles. This environment of the cell is made up of hydrophilic and hydrophobic regions which allows for the exchange of the above-mentioned molecules and ions. The hydrophilic regions of the cell are mainly on the inside and outside of the cell, while the hydrophobic regions are within the phospholipid bilayer of the cell membrane. The cell membrane consists of lipids and proteins which accounts for its hydrophobicity as a result of being non-polar substances. Therefore, in order for these molecules to participate in reactions, within the cell, they need to be able to cross this membrane layer to get into the cell. They accomplish this process of gaining access to the cell via: osmotic pressure, diffusion, concentration gradients, and membrane channels. Inside of the cell are extensive internal sub-cellular membrane-bounded compartments called organelles.

Organelles

- Centrosome - an associated pair of cylindrical shaped protein structures (centrioles) that organize microtubules and aid in forming the mitotic spindle during cell division in eukaryotes
- Cell membrane (plasma membrane) - the part of the cell which separates the cells from the outside environment and protects the cell, as well as regulating what goes in and out of the cell
- Cell wall - extra layer of protection and gives structural support (only found in plant cells)
- Chloroplast - key organelle for photosynthesis (only found in plant cells)
- Cilium - motile structure of eukaryotes having a cytoskeleton, the axoneme.
- Cytoplasm - contents of the main fluid-filled space inside cells, chemical reactions also happen in this jelly-like substance.
- Cytoskeleton - protein filaments inside cells (microfilaments, microtubules, and intermediate filaments)
- Endoplasmic reticulum (rough) - major site of membrane protein synthesis
- Endoplasmic reticulum (smooth) - major site of lipid synthesis
- Endosomes - vesicles that traffic membrane and intra and extra cellular contents for recycling or degradation by lysosomes
- Flagellum - motile structure of bacteria, archaea and eukaryotes
- Golgi apparatus - site of protein glycosylation in the endomembrane system

- Lipid bilayer - fundamental organizational structure of cell membranes
- Lysosome - acidic organelle that breaks down cellular waste products and debris into simple compounds (only found in animal cells)
- Microvilli - increases surface area for absorption of nutrients from surrounding medium
- Mitochondrion - major energy-producing organelle by releasing energy in the form of ATP
- Nucleus - contains chromosomes composed of DNA, the building block of life. Nuclear Architecture is important for dictating nuclear function.
- Organelle - term used for major subcellular structures
- Peroxisomes - a very small organelle that uses oxygen to breakdown and detoxify long fatty acids and other molecules
- Pili - also called fimbria is used for conjugation and sometimes movement
- Ribosome - RNA and protein complex required for protein synthesis in cells
- Starch grain - found in the cytoplasm of a typical plant cell, it stores chemical energy of the plant.
- Vacuole - contain cell sap (only found in plant cells)
- Vesicle - small membrane-bounded spheres inside cells

cell surface membrane protects the cell

Processes

Growth and Development

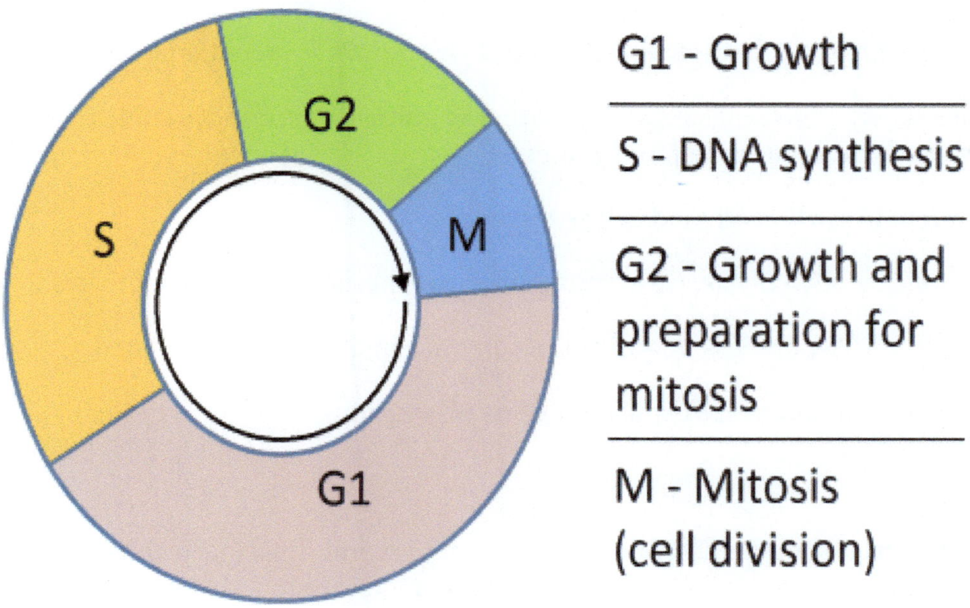

General concept of the cell cycle.

The growth process of the cell does not refer to the size of the cell, but instead the density of the number of cells present in the organism at a given time. Cell growth pertains to the increase in the number of cells present in an organism as it grows and develops; as the organism gets larger so too does the number of cells present. Cells are the foundation of all organisms, they are the fundamental unit of life. The growth and development of the cell are essential for the maintenance of the host, and survival of the organisms. For this process the cell goes through the steps of the cell cycle and development which involves cell growth, DNA replication, cell division, regeneration, specialization, and cell death. The cell cycle is divided into four distinct phases, G1, S, G2, and M. The G phases – which is the cell growth phase - makes up approximately 95% of the cycle. The proliferation of cells is instigated by progenitors, the cells then differentiate to become specialized, where specialized cells of the same type aggregate to form tissues, then organs and ultimately systems. The G phases along with the S phase – DNA replication, damage and repair - are considered to be the interphase portion of the cycle. While the M phase (mitosis and cytokinesis) is the cell division portion of the cycle. The cell cycle is regulated by a series of signalling factors and complexes such as CDK's, kinases, and p53. to name a few. When the cell has completed its growth process, and if it is found to be damaged or altered it undergoes cell death, either by apoptosis or necrosis, to eliminate the threat it cause to the organism's survival.

Other Cellular Processes

- Active transport and Passive transport - Movement of molecules into and out of cells.
- Autophagy - The process whereby cells "eat" their own internal components or microbial invaders.
- Adhesion - Holding together cells and tissues.
- Cell movement - Chemotaxis, contraction, cilia and flagella.
- Cell signaling - Regulation of cell behavior by signals from outside.
- Division - By which cells reproduce either by mitosis (to produce clones of the parent cell) or Meiosis (to produce haploid gametes)
- DNA repair - Cell death and cell senescence.
- Metabolism - Glycolysis, respiration, photosynthesis, and chemosynthesis.
- Signalling - The process by which the activities in the cell are regulated
- Transcription and mRNA splicing - Gene expression.

Techniques Used to Study Cells

Cells may be observed under the microscope, using several different techniques; these include optical microscopy, transmission electron microscopy, scanning electron microscopy, fluorescence microscopy, and confocal microscopy.

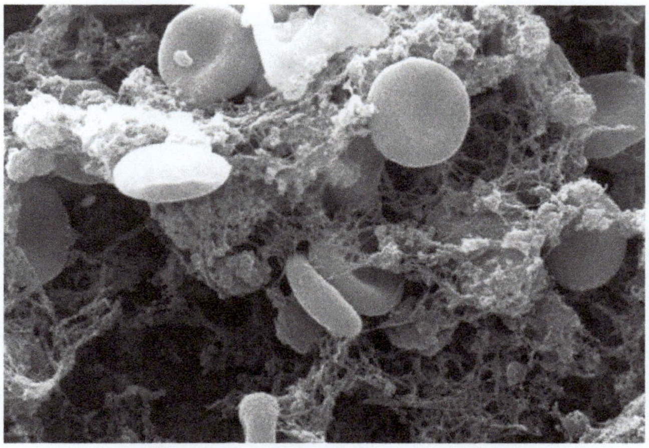

Electron micrograph of blood cells clotting.

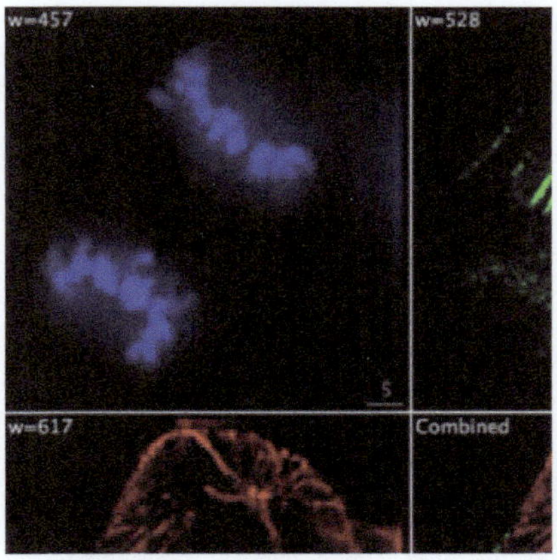

Cell division studied using fluorescence to stain specific structures

There are several different methods used in the study of cells:

- Cell culture is the basic technique of growing cells in a laboratory independent of an organism.
- Immunostaining, also known as immunohistochemistry, is a specialized histological method used to localize proteins in cells or tissue slices. Unlike regular histology, which uses stains to identify cells, cellular components or protein classes, immunostaining requires the reaction of an antibody directed against the protein of interest within the tissue or cell. Through the use of proper controls and published protocols (need to add reference links here), specificity of the antibody-antigen reaction can be achieved. Once this complex is formed, it is identified via either a "tag" attached directly to the antibody, or added in an additional technical step. Commonly used "tags" include fluorophores or enzymes. In the case of the former, detection of the location of the "immuno-stained" protein occurs via fluorescence microscopy. With an enzymatic tag, such as horse radish peroxidase, a chemical reaction is carried out that results in a dark color in the location of the protein of interest. This darkened pattern is then detected using light microscopy.
- Computational genomics is used to find patterns in genomic information
- DNA microarrays identify changes in transcript levels between different experimental conditions.

- Gene knockdown mutates a selected gene.
- In situ hybridization shows which cells are expressing a particular RNA transcript.
- PCR can be used to determine how many copies of a gene are present in a cell.
- Transfection introduces a new gene into a cell, usually an expression construct

Purification of cells and their parts Purification may be performed using the following methods:

- Cell fractionation
    - Release of cellular organelles by disruption of cells.
    - Separation of different organelles by centrifugation.
- Flow cytometry
- Immunoprecipitation
    - The binding of an antibody to a target protein
    - Collection of the target protein through elution
- Proteins extracted from cell membranes by detergents and salts or other kinds of chemicals.

Career and Related Fields

Practical job applications for a degree in Cell Molecular Biology includes the following.

• Agriculture • Bioinformatics & Medical Informatics • Biomedical Research • Biotechnology • Biotechnology Policy • Brewery Production • Business • Clinical Diagnostics • Colleges and University Teaching • Crop improvement • Elementary, Middle and High School Teaching • Environmental Technician/ Consultant • Food Production and Safety • Forensic Scientist • Geneticist	• Genetic Engineering • Government Jobs • Healthcare Industries • Health Policies • Health Law • Laboratory Technician • Medical Research • Medical School • Medical Technology • Microbiologist • Nutrition • Patent Law • Patient Care • Pest control • Pharmaceutical Science • Pharmaceutical Sales Representative • Physician Assistant • Product Quality and Safety	• Research Assistant • Research Coordinator • Research Scientist • Research Technician • Science Administration • Science Education • Science Journalism • Science Policy • Scientific Editing • Scientific Illustrator • Scientific Instrumentation Companies • Scientific Supply Representative • Teacher • Technical writing • Toxicology • Training and outreach in Industry and Government

References

- Cristianini, N. and Hahn, M. Introduction to Computational Genomics, Cambridge University Press, 2006. (ISBN 9780521671910 | ISBN 0-521-67191-4)

2

Types of Cell

Cells carry out different functions and are arranged according to the role that they play in an organism's lifetime. Cells can be classified as single-celled or multi-celled correlated to the evolutionary stage of the life form. Cell function also changes according to these differences.

Cell Type

A cell type is a classification used to distinguish between morphologically or phenotypically distinct cell forms within a species. A multicellular organism may contain a number of widely differing and specialised cell types, such as muscle cells and skin cells in humans, that differ both in appearance and function yet are genetically identical. Cells are able to be of the same genotype, but different cell type due to the differential regulation of the genes they contain. Classification of a specific cell type is often done through the use of microscopy and (such as those from the cluster of differentiation family that are commonly used for this purpose in immunology).

Animals have evolved a greater diversity of cell types in a multicellular body (100–150 different cell types), compared with 10–20 in plants, fungi, and protoctists.

Multicellular Organisms

All higher multicellular organisms contain cells specialised for different functions. Most distinct cell types arise from a single totipotent cell that differentiates into hundreds of different cell types during the course of development. Differentiation of cells is driven by different environmental cues (such as cell–cell interaction) and intrinsic differences (such as those caused by the uneven distribution of molecules during division). Multicellular organisms are composed of cells that fall into two fundamental types: germ cells and somatic cells. During development, somatic cells will become more specialized and form the three primary germ layers: ectoderm, mesoderm, and endoderm. After formation of the three germ layers, cells will continue to specialize until they reach a terminally differentiated state that is much more resistant to changes in cell type than its progenitors.

Humans

A list of cell types in the human body may include several hundred distinct types depending on the source.

Cells in humans derived from ectoderm		
Surface ectoderm	Skin	• Trichocyte • Keratinocyte
	Anterior pituitary	• Gonadotrope • Corticotrope • Thyrotrope • Somatotrope • Lactotroph
Neural crest	Peripheral nervous system	• Neuron • Glia Schwann cell • Satellite glial cell
	Neuroendocrine system	• Chromaffin cell • Parafollicular cell • Glomus cell
	Skin	• Melanocyte Nevus cell • Merkel cell
	Teeth	• Odontoblast • Cementoblast
	Eyes	• Corneal keratocyte
Neural tube	Central nervous system	• Neuron • Glia (Oligodendrocyte • Astrocyte)
	Ependyma	• Ependymocytes
	Pineal gland	• Pinealocyte

Cells in humans derived from mesoderm			
Paraxial	Mesenchymal stem cell	Osteochondroprogenitor cell	• Bone (Osteoblast → Osteocyte) Cartilage • (Chondroblast → Chondrocyte)
		Myofibroblast	• Fat Lipoblast → Adipocyte • Muscle Myoblast → Myocyte Myosatellite cell Tendon cell Cardiac muscle cell • Other Fibroblast → Fibrocyte
	Other	• Digestive system → Interstitial cell of Cajal	
Intermediate	Renal stem cell	• Angioblast → Endothelial cell • Mesangial cell Intraglomerular Extraglomerular • Juxtaglomerular cell • Macula densa cell • Stromal cell → Interstitial cell → Telocytes • Simple epithelial cell → Podocyte • Kidney proximal tubule brush border cell	
Lateral plate/ hemangioblast	Hematopoietic stem cell	• Lymphoid • Lymphoblast • lymphocytes • Myeloid • CFU-GEMM	
	Circulatory system	• Endothelial progenitor cell • Endothelial colony forming cell • Endothelial stem cell • Angioblast/Mesoangioblast • Pericyte • Mural cell	

Prokaryote

A prokaryote is a single-celled organism that lacks a membrane-bound nucleus (karyon), mitochondria, or any other membrane-bound organelle. Prokaryotes can be divided into two domains, Archaea and Bacteria. In contrast, species with nuclei and organelles are placed in the domain Eukaryota.

In the prokaryotes, all the intracellular water-soluble components (proteins, DNA and metabolites) are located together in the cytoplasm enclosed by the cell membrane, rather than in separate cellular compartments. Bacteria, however, do possess protein-based bacterial microcompartments, which are thought to act as primitive organelles enclosed in protein shells. Some prokaryotes, such as cyanobacteria may form large colonies. Others, such as myxobacteria, have multicellular stages in their life cycles.

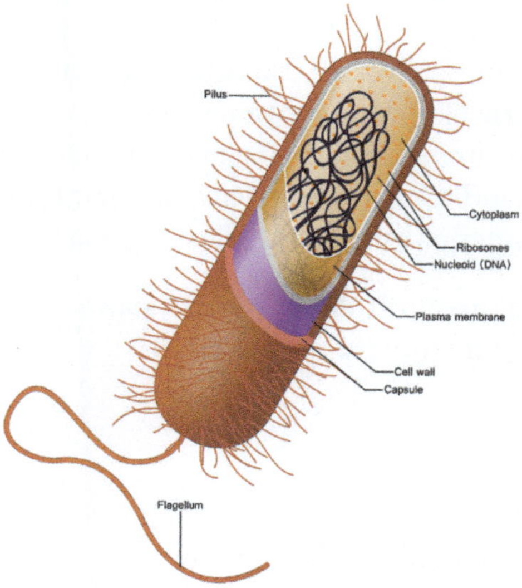

A diagram of a typical prokaryotic bacteria cell.

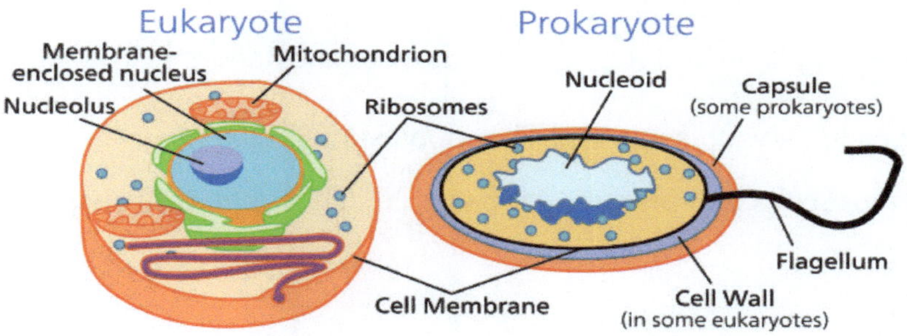

Comparison of Eukaryotes vs. Prokaryotes

Molecular studies have provided insight into the evolution and interrelationships of the three domains of biological species. Eukaryotes are organisms, including humans, whose cells have a well defined membrane-bound nucleus (containing chromosomal DNA) and organelles. The division between prokaryotes and eukaryotes reflects the existence of two very different levels of cellular organization. Distinctive types of prokaryotes include extremophiles and methanogens; these are common in some extreme environments.

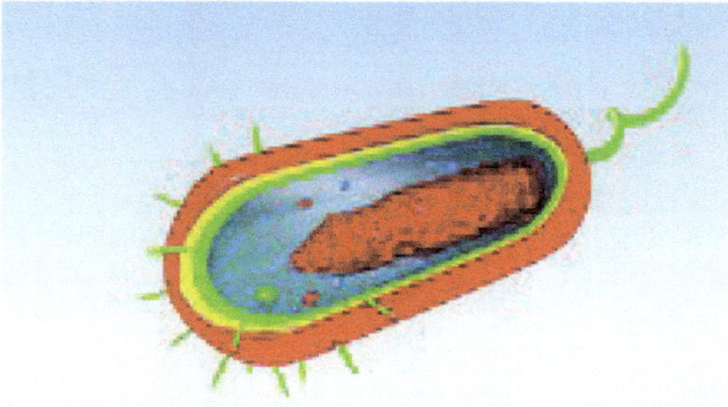

3D animation of a prokaryotic cell that shows all the elements that compose it

Structure

Prokaryotes have a prokaryotic cytoskeleton, albeit more primitive than that of the eukaryotes. Besides homologues of actin and tubulin (MreB and FtsZ), the helically arranged building-block of the flagellum, flagellin, is one of the most significant cytoskeletal proteins of bacteria, as it provides structural backgrounds of chemotaxis, the basic cell physiological response of bacteria. At least some prokaryotes also contain intracellular structures that can be seen as primitive organelles. Membranous organelles (or intracellular membranes) are known in some groups of prokaryotes, such as vacuoles or membrane systems devoted to special metabolic properties, such as photosynthesis or chemolithotrophy. In addition, some species also contain carbohydrate-enclosed microcompartments, which have distinct physiological roles (e.g. carboxysomes or gas vacuoles).

Most prokaryotes are between 1 μm and 10 μm, but they can vary in size from 0.2 μm (*Mycoplasma genitalium*) to 750 μm (*Thiomargarita namibiensis*).

Prokaryotic Cell Structure

Flagellum (only in some types of prokaryotes)

Long, whip-like protrusion that aids cellular locomotion.

Cell Membrane

Surrounds the cell's cytoplasm and regulates the flow of substances in and out of the cell.

Cell wall (except genera Mycoplasma and Thermoplasma)

Outer covering of most cells that protects the bacterial cell and gives it shape.

Cytoplasm

A gel-like substance composed mainly of water that also contains enzymes, salts, cell components, and various organic molecules.

Ribosome

Cell structures responsible for protein production.

Nucleoid

Area of the cytoplasm that contains the single bacterial DNA molecule.

Glycocalyx (only in some types of prokaryotes)

A glycoprotein-polysaccharide covering that surrounds the cell membranes.

Inclusions.

Morphology

Prokaryotic cells have various shapes; the four basic shapes of bacteria are:

- Cocci – spherical
- Bacilli – rod-shaped
- Spirochaete – spiral-shaped
- Vibrio – comma-shaped

The archaeon Haloquadratum has flat square-shaped cells.

Reproduction

Bacteria and archaea reproduce through asexual reproduction, usually by binary fission. Genetic exchange and recombination still occur, but this is a form of horizontal gene transfer and is not a replicative process, simply involving the transference of DNA between two cells, as in bacterial conjugation.

DNA Transfer

DNA transfer between prokaryotic cells occurs in bacteria and archaea, although it has been mainly studied in bacteria. In bacteria, gene transfer occurs by three processes. These are (1) bacterial virus (bacteriophage)-mediated transduction, (2) plasmid-mediated conjugation, and (3) natural transformation. Transduction of bacterial genes by bacteriophage appears to reflect an occasional error during intracellular assembly of virus particles, rather than an adaptation of the host bacteria. The transfer of bacterial DNA is under the control of the bacteriophage's genes rather than bacterial genes. Conjugation in the well-studied *E. coli* system is controlled by plasmid genes, and is an adaptation for distributing copies of a plasmid from one bacterial host to another. Infrequently during this process, a plasmid may integrate into the host bacterial chromosome, and subsequently transfer part of the host bacterial DNA to another bacterium. Plasmid mediated transfer of host bacterial DNA (conjugation) also appears to be an accidental process rather than a bacterial adaptation.

Natural bacterial transformation involves the transfer of DNA from one bacterium to another through the intervening medium. Unlike transduction and conjugation, transformation is clearly a bacterial adaptation for DNA transfer, because it depends on numerous bacterial gene products that specifically interact to perform this complex process. For a bacterium to bind, take up and recombine donor DNA into its own chromosome, it must first enter a special physiological state called competence. About 40 genes are required in *Bacillus subtilis* for the development of competence. The length of DNA transferred during *B. subtilis* transformation can be as much as a third to the whole chromosome. Transformation is a common mode of DNA transfer, and 67 prokaryotic species are thus far known to be naturally competent for transformation.

Among archaea, *Halobacterium volcanii* forms cytoplasmic bridges between cells that appear to be used for transfer of DNA from one cell to another. Another archaeon, *Sulfolobus solfataricus*, transfers DNA between cells by direct contact. Frols et al. found that exposure of *S. solfataricus* to

DNA damaging agents induces cellular aggregation, and suggested that cellular aggregation may enhance DNA transfer among cells to provide increased repair of damaged DNA via homologous recombination.

Sociality

While prokaryotes are considered strictly unicellular, most can form stable aggregate communities. When such communities are encased in a stabilizing polymer matrix ("slime"), they may be called "biofilms". Cells in biofilms often show distinct patterns of gene expression (phenotypic differentiation) in time and space. Also, as with multicellular eukaryotes, these changes in expression often appear to result from cell-to-cell signaling, a phenomenon known as quorum sensing.

Biofilms may be highly heterogeneous and structurally complex and may attach to solid surfaces, or exist at liquid-air interfaces, or potentially even liquid-liquid interfaces. Bacterial biofilms are often made up of microcolonies (approximately dome-shaped masses of bacteria and matrix) separated by "voids" through which the medium (e.g., water) may flow relatively uninhibited. The microcolonies may join together above the substratum to form a continuous layer, closing the network of channels separating microcolonies. This structural complexity—combined with observations that oxygen limitation (a ubiquitous challenge for anything growing in size beyond the scale of diffusion) is at least partially eased by movement of medium throughout the biofilm—has led some to speculate that this may constitute a circulatory system and many researchers have started calling prokaryotic communities multicellular (for example). Differential cell expression, collective behavior, signaling, programmed cell death, and (in some cases) discrete biological dispersal events all seem to point in this direction. However, these colonies are seldom if ever founded by a single founder (in the way that animals and plants are founded by single cells), which presents a number of theoretical issues. Most explanations of co-operation and the evolution of multicellularity have focused on high relatedness between members of a group (or colony, or whole organism). If a copy of a gene is present in all members of a group, behaviors that promote cooperation between members may permit those members to have (on average) greater fitness than a similar group of selfish individuals.

Should these instances of prokaryotic sociality prove to be the rule rather than the exception, it would have serious implications for the way we view prokaryotes in general, and the way we deal with them in medicine. Bacterial biofilms may be 100 times more resistant to antibiotics than free-living unicells and may be nearly impossible to remove from surfaces once they have colonized them. Other aspects of bacterial cooperation—such as bacterial conjugation and quorum-sensing-mediated pathogenicity, present additional challenges to researchers and medical professionals seeking to treat the associated diseases.

Environment

Prokaryotes have diversified greatly throughout their long existence. The metabolism of prokaryotes is far more varied than that of eukaryotes, leading to many highly distinct prokaryotic types. For example, in addition to using photosynthesis or organic compounds for energy, as eukaryotes do, prokaryotes may obtain energy from inorganic compounds such as hydrogen sulfide. This enables prokaryotes to thrive in harsh environments as cold as the snow surface of Antarctica, studied in cryobiology or as hot as undersea hydrothermal vents and land-based hot springs.

Prokaryotes live in nearly all environments on Earth. Some archaea and bacteria thrive in harsh conditions, such as high temperatures (thermophiles) or high salinity (halophiles). Organisms such as these are referred to as extremophiles. Many archaea grow as plankton in the oceans. Symbiotic prokaryotes live in or on the bodies of other organisms, including humans.

Classification

In 1977, Carl Woese proposed dividing prokaryotes into the Bacteria and Archaea (originally Eubacteria and Archaebacteria) because of the major differences in the structure and genetics between the two groups of organisms. Archaea were originally thought to be extremophiles, living only in inhospitable conditions such as extremes of temperature, pH, and radiation but have since been found in all types of habitats. The resulting arrangement of Eukaryota (also called "Eukarya"), Bacteria, and Archaea is called the three-domain system, replacing the traditional two-empire system.

One criticism of the dichotomous prokaryote-eukaryote distinction points out that the word "prokaryote" is based on what these organisms are not (they are not eukaryotic), rather than what they are (either archaea or bacteria).

Evolution of Prokaryotes

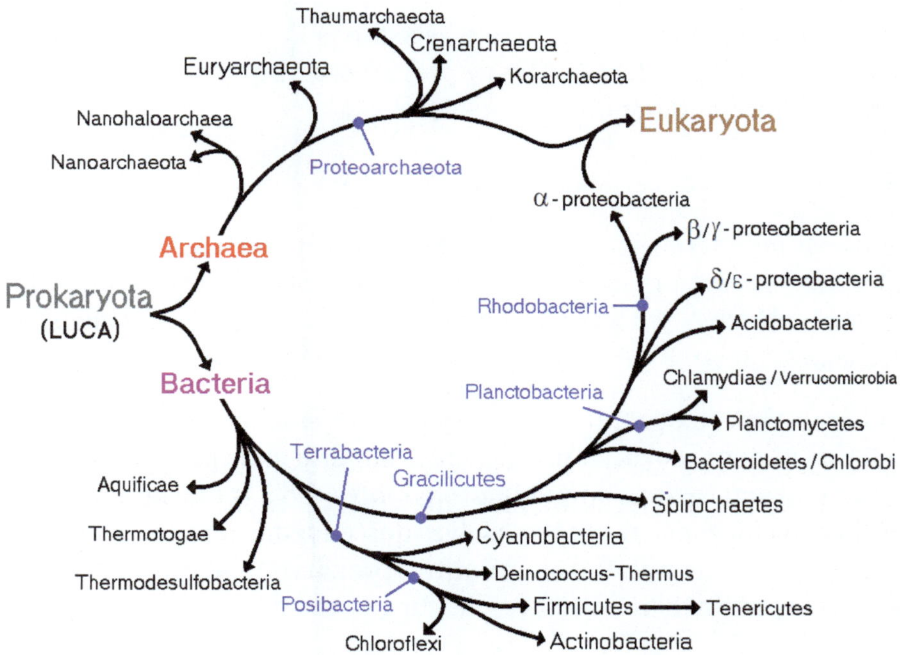

Phylogenetic ring showing the diversity of prokaryotes, and symbiogenetic origins of eukaryotes

The current model of the evolution of the first living organisms is that these were some form of prokaryotes, which may have evolved out of protocells. In general, the eukaryotes are thought to have evolved later in the history of life.

However, some authors have questioned this conclusion, arguing that the current set of prokaryotic species may have evolved from more complex eukaryotic ancestors through a process of simplification.

Others have argued that the three domains of life arose simultaneously, from a set of varied cells that formed a single gene pool.

This controversy was summarized in 2005:

There is no consensus among biologists concerning the position of the eukaryotes in the overall scheme of cell evolution. Current opinions on the origin and position of eukaryotes span a broad spectrum including the views that eukaryotes arose first in evolution and that prokaryotes descend from them, that eukaryotes arose contemporaneously with eubacteria and archeabacteria and hence represent a primary line of descent of equal age and rank as the prokaryotes, that eukaryotes arose through a symbiotic event entailing an endosymbiotic origin of the nucleus, that eukaryotes arose without endosymbiosis, and that eukaryotes arose through a symbiotic event entailing a simultaneous endosymbiotic origin of the flagellum and the nucleus, in addition to many other models, which have been reviewed and summarized elsewhere.

The oldest known fossilized prokaryotes were laid down approximately 3.5 billion years ago, only about 1 billion years after the formation of the Earth's crust. Eukaryotes only appear in the fossil record later, and may have formed from endosymbiosis of multiple prokaryote ancestors. The oldest known fossil eukaryotes are about 1.7 billion years old. However, some genetic evidence suggests eukaryotes appeared as early as 3 billion years ago.

While Earth is the only place in the universe where life is known to exist, some have suggested that there is evidence on Mars of fossil or living prokaryotes; but this possibility remains the subject of considerable debate and skepticism.

Relationship to Eukaryotes

The division between prokaryotes and eukaryotes is usually considered the most important distinction or difference among organisms. The distinction is that eukaryotic cells have a "true" nucleus containing their DNA, whereas prokaryotic cells do not have a nucleus. Both eukaryotes and prokaryotes contain large RNA/protein structures called ribosomes, which produce protein.

Another difference is that ribosomes in prokaryotes are smaller than in eukaryotes. However, two organelles found in many eukaryotic cells, mitochondria and chloroplasts, contain ribosomes similar in size and makeup to those found in prokaryotes. This is one of many pieces of evidence that mitochondria and chloroplasts are themselves descended from free-living bacteria. This theory holds that early eukaryotic cells took in primitive prokaryotic cells by phagocytosis and adapted themselves to incorporate their structures, leading to the mitochondria we see today.

The genome in a prokaryote is held within a DNA/protein complex in the cytosol called the nucleoid, which lacks a nuclear envelope. The complex contains a single, cyclic, double-stranded molecule of stable chromosomal DNA, in contrast to the multiple linear, compact, highly organized chromosomes found in eukaryotic cells. In addition, many important genes of prokaryotes are stored in separate circular DNA structures called plasmids.

Prokaryotes lack mitochondria and chloroplasts. Instead, processes such as oxidative phosphorylation and photosynthesis take place across the prokaryotic cell membrane. However, prokaryotes do possess some internal structures, such as prokaryotic cytoskeletons. It has been suggested that the bacterial order Planctomycetes have a membrane around their nucleoid and contain other membrane-bound cellular structures. However, further investigation revealed that Planctomycetes cells are not compartmentalized or nucleated and like the other bacterial membrane systems are all interconnected.

Prokaryotic cells are usually much smaller than eukaryotic cells. Therefore, prokaryotes have a larger surface-area-to-volume ratio, giving them a higher metabolic rate, a higher growth rate, and as a consequence, a shorter generation time than eukaryotes.

Eukaryote

A eukaryote is any organism whose cells contain a nucleus and other organelles enclosed within membranes.

Eukaryotes belong to the taxon Eukarya or Eukaryota. The defining feature that sets eukaryotic cells apart from prokaryotic cells (Bacteria and Archaea) is that they have membrane-bound organelles, especially the nucleus, which contains the genetic material, and is enclosed by the nuclear envelope. Eukaryotic cells also contain other membrane-bound organelles such as mitochondria and the Golgi apparatus. In addition, plants and algae contain chloroplasts. Eukaryotic organisms may be unicellular, or multicellular. Only eukaryotes form multicellular organisms consisting of many kinds of tissue made up of different cell types.

Eukaryotes can reproduce both asexually through mitosis and sexually through meiosis and gamete fusion. In mitosis, one cell divides to produce two genetically identical cells. In meiosis, DNA replication is followed by two rounds of cell division to produce four daughter cells each with half the number of chromosomes as the original parent cell (haploid cells). These act as sex cells (gametes – each gamete has just one complement of chromosomes, each a unique mix of the corresponding pair of parental chromosomes) resulting from genetic recombination during meiosis.

The domain Eukaryota appears to be monophyletic, and so makes up one of the three domains of life. The two other domains, Bacteria and Archaea, are prokaryotes and have none of the above features. Eukaryotes represent a tiny minority of all living things. However, due to their much larger size, eukaryotes' collective worldwide biomass is estimated at about equal to that of prokaryotes. Eukaryotes first developed approximately 1.6–2.1 billion years ago (during the proterozoic era).

Cell Features

Eukaryotic cells are typically much larger than those of prokaryotes. They have a variety of internal membrane-bound structures, called organelles, and a cytoskeleton composed of microtubules, microfilaments, and intermediate filaments, which play an important role in defining the cell's organization and shape. Eukaryotic DNA is divided into several linear bundles called chromosomes, which are separated by a microtubular spindle during nuclear division.

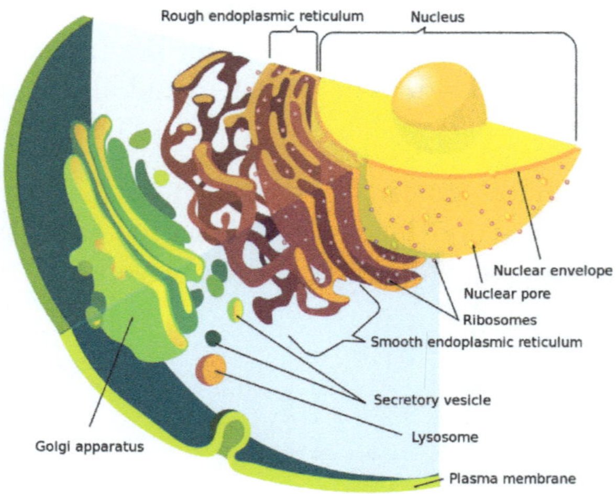

Detail of the endomembrane system and its components

A 3D rendering of an animal cell cut in half.

Internal Membrane

Eukaryote cells include a variety of membrane-bound structures, collectively referred to as the endomembrane system. Simple compartments, called vesicles or vacuoles, can form by budding off other membranes. Many cells ingest food and other materials through a process of endocytosis, where the outer membrane invaginates and then pinches off to form a vesicle. It is probable that most other membrane-bound organelles are ultimately derived from such vesicles. Alternatively some products produced by the cell can leave in a vesicle through exocytosis.

The nucleus is surrounded by a double membrane (commonly referred to as a nuclear envelope), with pores that allow material to move in and out. Various tube- and sheet-like extensions of the nuclear membrane form what is called the endoplasmic reticulum or ER, which is involved in protein transport and maturation. It includes the rough ER where ribosomes are attached to synthesize proteins, which enter the interior space or lumen. Subsequently, they generally enter vesicles, which bud off from the smooth ER. In most eukaryotes, these protein-carrying vesicles are released and further modified in stacks of flattened vesicles, called Golgi bodies or dictyosomes.

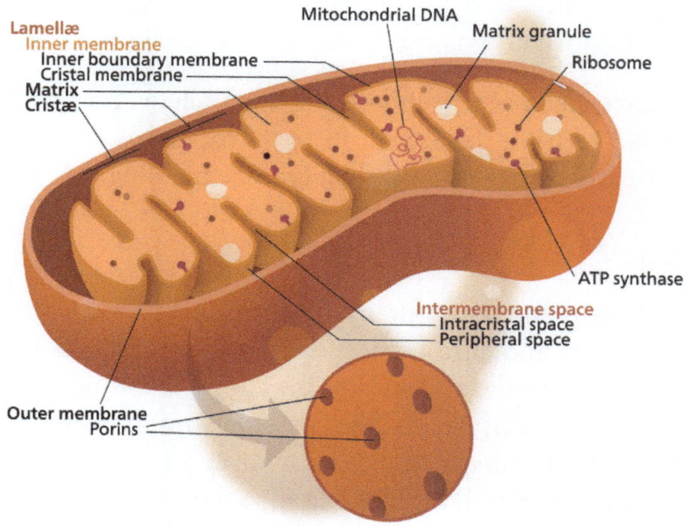

Simplified structure of a mitochondrion

Vesicles may be specialized for various purposes. For instance, lysosomes contain digestive enzymes that break down the contents of food vacuoles, and peroxisomes are used to break down peroxide, which is toxic otherwise. Many protozoa have contractile vacuoles, which collect and expel excess water, and extrusomes, which expel material used to deflect predators or capture prey. In higher plants, most of a cell's volume is taken up by a central vacuole, which primarily maintains its osmotic pressure.

Mitochondria and Plastids

Mitochondria are organelles found in nearly all eukaryotes. They are the powerbank or powerhouse of the cell as they provide energy to the cell by converting ingested sugars into ATP. They are surrounded by two membranes (each a phospholipid bi-layer), the inner of which is folded into invaginations called cristae, where aerobic respiration takes place. Mitochondria contain their own DNA. They are now generally held to have developed from endosymbiotic prokaryotes, probably proteobacteria. Protozoa and microbes that lack mitochondria, such as the amoebozoan Pelomyxa and metamonads such as Giardia and Trichomonas, have usually been found to contain mitochondrion-derived organelles, such as hydrogenosomes and mitosomes, and thus probably lost the mitochondria secondarily.

In 2016, *Monocercomonoides*, a metamonad flagellate which resides in the intestines of the chinchilla, has been found to lack mitochondria entirely. *Monocercomonoides* obtains its energy by enzymatic action on nutrients absorbed from the environment. It has also acquired, by lateral gene transfer, a cytosolic sulphur mobilisation system which provides the clusters of iron and sulfur required for protein synthesis. The normal mitochondrial iron-sulphur cluster pathway is considered to have been lost secondarily.

Plants and various groups of algae also have plastids. Plastids have their own DNA and are developed from endosymbionts, in this case cyanobacteria. They usually take the form of chloroplasts, which like cyanobacteria contain chlorophyll and produce organic compounds (such as glucose) through photosynthesis. Others are involved in storing food. Although plastids probably had a single origin, not all plastid-containing groups are closely related. Instead, some eukaryotes have obtained them from others through secondary endosymbiosis or ingestion.

Endosymbiotic origins have also been proposed for the nucleus, for which see below, and for eukaryotic flagella, supposed to have developed from spirochaetes.This is not generally accepted, both from a lack of cytological evidence and difficulty in reconciling this with cellular reproduction.

Cytoskeletal Structures

Many eukaryotes have long slender motile cytoplasmic projections, called flagella, or similar structures called cilia. Flagella and cilia are sometimes referred to as undulipodia, and are variously involved in movement, feeding, and sensation. They are composed mainly of tubulin. These are entirely distinct from prokaryotic flagellae. They are supported by a bundle of microtubules arising from a basal body, also called a kinetosome or centriole, characteristically arranged as nine doublets surrounding two singlets. Flagella also may have hairs, or mastigonemes, and scales connecting membranes and internal rods. Their interior is continuous with the cell's cytoplasm.

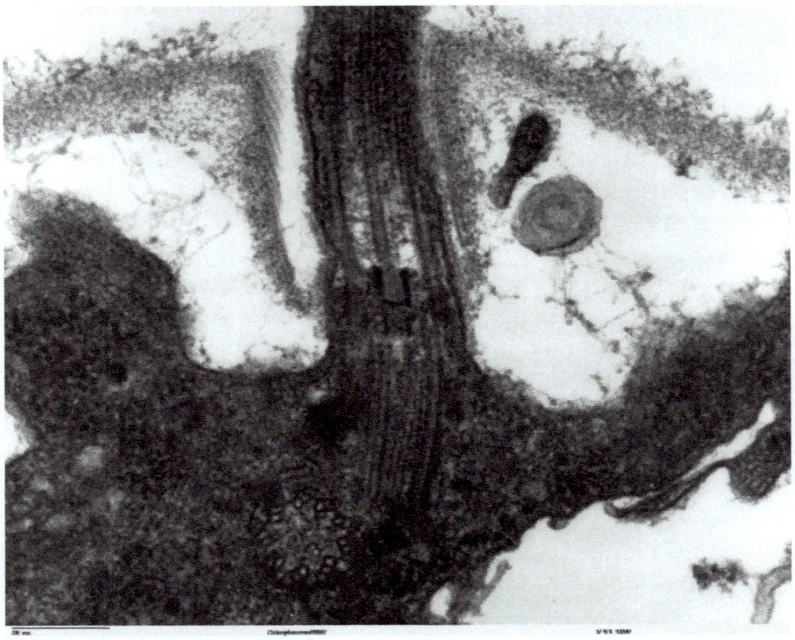

Longitudinal section through the flagellum of *Chlamydomonas reinhardtii*

Microfilamental structures composed of actin and actin binding proteins, e.g., α-actinin, fimbrin, filamin are present in submembraneous cortical layers and bundles, as well. Motor proteins of microtubules, e.g., dynein or kinesin and actin, e.g., myosins provide dynamic character of the network.

Centrioles are often present even in cells and groups that do not have flagella, but conifers and flowering plants have neither. They generally occur in groups of one or two, called kinetids, that give rise to various microtubular roots. These form a primary component of the cytoskeletal structure, and are often assembled over the course of several cell divisions, with one flagellum retained from the parent and the other derived from it. Centrioles may also be associated in the formation of a spindle during nuclear division.

The significance of cytoskeletal structures is underlined in the determination of shape of the cells, as well as their being essential components of migratory responses like chemotaxis and chemokinesis. Some protists have various other microtubule-supported organelles. These include the radiolaria and heliozoa, which produce axopodia used in flotation or to capture prey, and the haptophytes, which have a peculiar flagellum-like organelle called the haptonema.

Cell Wall

The cells of plants, fungi, and most chromalveolates have a cell wall, a layer outside the cell membrane, providing the cell with structural support,protection, and a filtering mechanism. The cell wall also prevents over-expansion when water enters the cell.

The major polysaccharides making up the primary cell wall of land plants are cellulose, hemicellulose, and pectin. The cellulose microfibrils are linked via hemicellulosic tethers to form the cellulose-hemicellulose network, which is embedded in the pectin matrix. The most common hemicellulose in the primary cell wall is xyloglucan.

Differences Among Eukaryotic Cells

There are many different types of eukaryotic cells, though animals and plants are the most familiar eukaryotes, and thus provide an excellent starting point for understanding eukaryotic structure. Fungi and many protists have some substantial differences, however.

Animal Cell

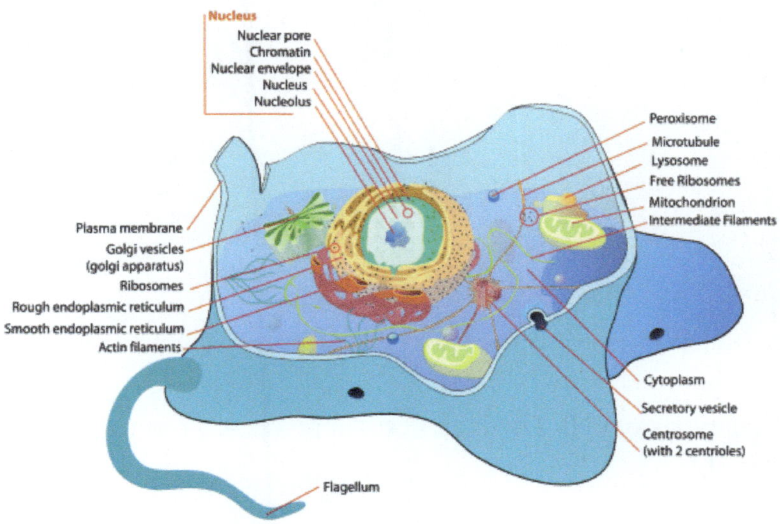

Structure of a typical animal cell

All animals consist of eukaryotic cells. Animal cells are distinct from those of other eukaryotes, most notably plants, as they lack cell walls and chloroplasts and have smaller vacuoles. Due to the lack of a cell wall, animal cells can adopt a variety of shapes. A phagocytic cell can even engulf other structures.

3D Simulation of animal cell

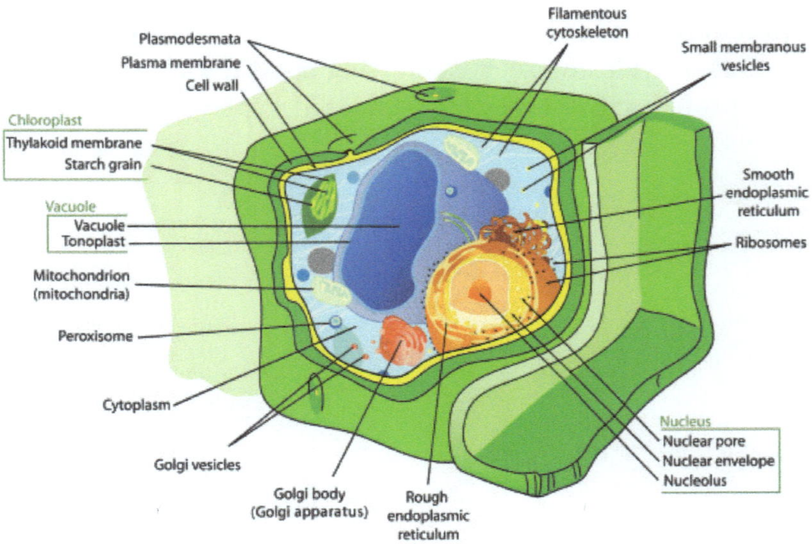

Structure of a typical plant cell

There are many other types of cell. For instance, there are approximately 210 distinct cell types in the adult human body.

Plant Cell

Plant cells are quite different from the cells of the other eukaryotic organisms. Their distinctive features are:

- A large central vacuole (enclosed by a membrane, the tonoplast), which maintains the cell's turgor and controls movement of molecules between the cytosol and sap

- A primary cell wall containing cellulose, hemicellulose and pectin, deposited by the protoplast on the outside of the cell membrane; this contrasts with the cell walls of fungi, which contain chitin, and the cell envelopes of prokaryotes, in which peptidoglycans are the main structural molecules
- The plasmodesmata, linking pores in the cell wall that allow each plant cell to communicate with other adjacent cells; this is different from the functionally analogous system of gap junctions between animal cells.
- Plastids, especially chloroplasts that contain chlorophyll, the pigment that gives plants their green color and allows them to perform photosynthesis
- Bryophytes and seedless vascular plants lack flagellae and centrioles except in the sperm cells. Sperm of cycads and *Ginkgo* are large, complex cells that swim with hundreds to thousands of flagellae.
- Conifers (Pinophyta) and flowering plants (Angiospermae) lack the flagellae and centrioles that are present in animal cells.

Fungal Cell

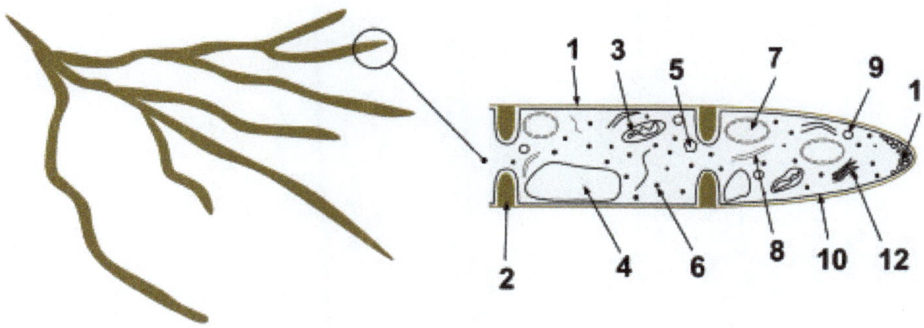

Fungal Hyphae Cells

1- Hyphal wall 2- Septum 3- Mitochondrion 4- Vacuole 5- Ergosterol crystal 6- Ribosome 7- Nucleus 8- Endoplasmic reticulum 9- Lipid body 10- Plasma membrane 11- Spitzenkörper 12- Golgi apparatus

Fungal cells are most similar to animal cells, with the following exceptions:

- A cell wall that contains chitin
- Less definition between cells; the hyphae of higher fungi have porous partitions called septa, which allow the passage of cytoplasm, organelles, and, sometimes, nuclei. Primitive fungi have few or no septa, so each organism is essentially a giant multinucleate supercell; these fungi are described as coenocytic.
- Only the most primitive fungi, chytrids, have flagella.

Other Eukaryotic Cells

Eukaryotes are a very diverse group, and their cell structures are equally diverse. Many have cell walls; many do not. Many have chloroplasts, derived from primary, secondary, or even tertiary endosymbiosis; and many do not. Some groups have unique structures, such as the cyanelles of the glaucophytes, the haptonema of the haptophytes, or the ejectisomes of the cryptomonads. Other structures, such as pseudopods, are found in various eukaryote groups in different forms, such as the lobose amoebozoans or the reticulose foraminiferans.

Reproduction

Cell division generally takes place asexually by mitosis, a process that allows each daughter nucleus to receive one copy of each chromosome. In most eukaryotes, there is also a process of sexual reproduction, typically involving an alternation between haploid generations, wherein only one copy of each chromosome is present, and diploid generations, wherein two copies of each chromosome are present, occurring through meiosis. There is considerable variation in this pattern.

Eukaryotes have a smaller surface area to volume ratio than prokaryotes, and thus have lower metabolic rates and longer generation times. In some multicellular organisms, cells specialized for metabolism will have enlarged surface areas, such as intestinal vili.

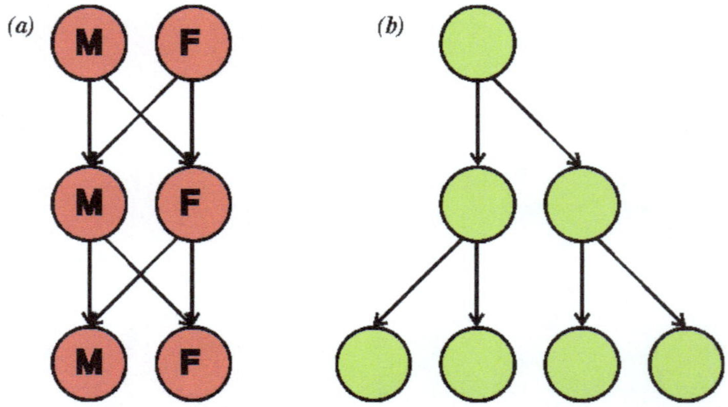

This diagram illustrates the *twofold cost of sex*. If each individual were to contribute to the same number of offspring (two), *(a)* the sexual population remains the same size each generation, where the *(b)* asexual population doubles in size each generation.

The evolution of sexual reproduction may be a primordial and fundamental characteristic of eukaryotes. Based on a phylogenetic analysis, Dacks and Roger proposed that facultative sex was present in the common ancestor of all eukaryotes. A core set of genes that function in meiosis is present in both *Trichomonas vaginalis* and *Giardia intestinalis*, two organisms previously thought to be asexual. Since these two species are descendants of lineages that diverged early from the eukaryotic evolutionary tree, it was inferred that core meiotic genes, and hence sex, were likely present in a common ancestor of all eukaryotes. Other studies on eukaryotic species once thought to be asexual have revealed evidence for a sexual cycle. For instance, parasitic protozoa of the genus Leishmania have recently been shown to have a sexual cycle. Also, evidence now indicates that amoeba, that were previously regarded as asexual, are anciently sexual and that the majority of present-day asexual groups likely arose recently and independently.

Classification

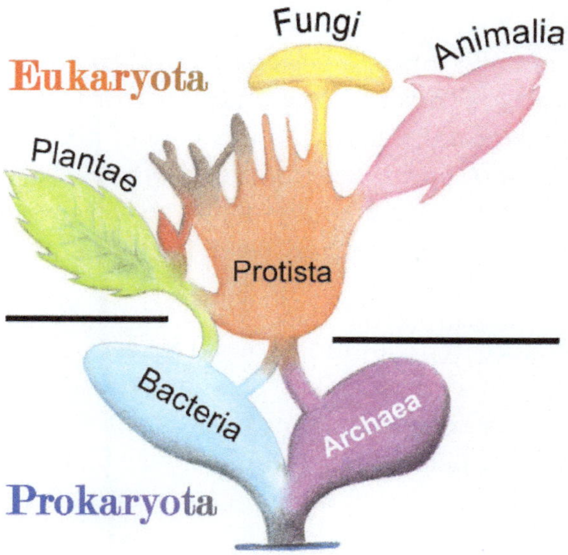

Phylogenetic and symbiogenetic tree of living organisms, showing the origins of eukaryotes & prokaryotes

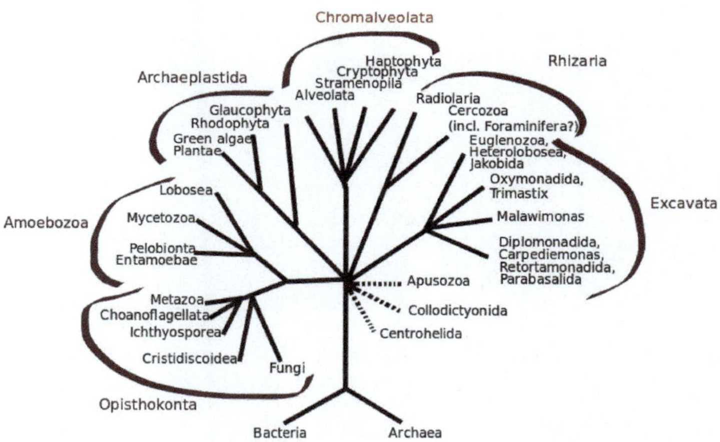

One hypothesis of eukaryotic relationships. The Opisthokonta group includes both animals (Metazoa) and fungi. Plants (Plantae) are placed in Archaeplastida.

In antiquity, the two clades of animals and plants were recognized. They were given the taxonomic rank of Kingdom by Linnaeus. Though he included the fungi with plants with some reservations, it was later realized that they are quite distinct and warrant a separate kingdom, the composition of which was not entirely clear until the 1980s. The various single-cell eukaryotes were originally placed with plants or animals when they became known. In 1830, the German biologist Georg A. Goldfuss coined the word *protozoa* to refer to organisms such as ciliates, and this group was expanded until it encompassed all single-celled eukaryotes, and given their own kingdom, the Protista, by Ernst Haeckel in 1866. The eukaryotes thus came to be composed of four kingdoms:

- Kingdom Protista
- Kingdom Plantae

- Kingdom Fungi
- Kingdom Animalia

The protists were understood to be "primitive forms", and thus an evolutionary grade, united by their primitive unicellular nature. The disentanglement of the deep splits in the tree of life only really got going with DNA sequencing, leading to a system of domains rather than kingdoms as top level rank being put forward by Carl Woese, uniting all the eukaryote kingdoms under the eukaryote domain. At the same time, work on the protist tree intensified, and is still actively going on today. Several alternative classifications have been forwarded, though there is no consensus in the field.

A classification produced in 2005 for the International Society of Protistologists, which reflected the consensus of the time, divided the eukaryotes into six supposedly monophyletic 'supergroups'. However, in the same year (2005), doubts were expressed as to whether some of these supergroups were monophyletic, particularly the Chromalveolata, and a review in 2006 noted the lack of evidence for several of the supposed six supergroups. A revised classification in 2012 recognizes five supergroups.

Archaeplastida (or Primoplantae)	Land plants, green algae, red algae, and glaucophytes
SAR supergroup	Stramenopiles (brown algae, diatoms, etc.), Alveolata, and Rhizaria (Foraminifera, Radiolaria, and various other amoeboid protozoa).
Excavata	Various flagellate protozoa
Amoebozoa	Most lobose amoeboids and slime molds
Opisthokonta	Animals, fungi, choanoflagellates, etc.

There are also smaller groups of eukaryotes whose position is uncertain or seems to fall outside the major groups. In particular, Haptophyta, Cryptophyta, Centrohelida, Telonemia, Picozoa, Apusomonadida, Ancyromonadida, Breviatea, and the genus *Collodictyon*. Overall, it seems that, although progress has been made, there are still very significant uncertainties in the evolutionary history and classification of eukaryotes. As Roger & Simpson said in 2009 "with the current pace of change in our understanding of the eukaryote tree of life, we should proceed with caution."

In an article published in *Nature Microbiology* in April 2016 the authors, "reinforced once again that the life we see around us – plants, animals, humans and other so-called eukaryotes – represent a tiny percentage of the world's biodiversity." They classified eukaryote "based on the inheritance of their information systems as opposed to lipid or other cellular structures." Jillian F. Banfield of the University of California, Berkeley and fellow scientists used a super computer to generate a diagram of a new tree of life based on DNA from 3000 species including 2,072 known species and 1,011 newly reported microbial organisms, whose DNA they had gathered from diverse environments. As the capacity to sequence DNA became easier, Banfield and team were able to do metagenomic sequencing—"sequencing whole communities of organisms at once and picking out the individual groups based on their genes alone."

Phylogeny

Life Timeline

The rRNA trees constructed during the 1980s and 1990s left most eukaryotes in an unresolved "crown" group (not technically a true crown), which was usually divided by the form of the mitochondrial cristae. The few groups that lack mitochondria branched separately, and so the absence was believed to be primitive; but this is now considered an artifact of long-branch attraction, and they are known to have lost them secondarily.

As of 2011, there is widespread agreement that the Rhizaria belong with the Stramenopiles and the Alveolata, in a clade dubbed the SAR supergroup, so that Rhizaria is not one of the main eukaryote groups; also that the Amoebozoa and Opisthokonta are each monophyletic and form a clade, often called the unikonts. Beyond this, there does not appear to be a consensus.

It has been estimated that there may be 75 distinct lineages of eukaryotes. Most of these lineages are protists.

The known eukaryote genome sizes vary from 8.2 megabases (Mb) in *Babesia bovis* to 112,000–220,050 Mb in the dinoflagellate *Prorocentrum micans*, suggesting that the genome of the ancestral eukaryote has undergone considerable variation during its evolution. The last common ancestor of all eukaryotes is believed to have been a phagotrophic protist with a nucleus, at least one centriole and cilium, facultatively aerobic mitochondria, sex (meiosis and syngamy), a dormant cyst with a cell wall of chitin and/or cellulose and peroxisomes. Later endosymbiosis led to the spread of plastids in some lineages.

Five Supergroups

A global tree of eukaryotes from a consensus of phylogenetic evidence (in particular, phylogenomics), rare genomic signatures, and morphological characteristics is presented in Adl *et al.* 2012 and Burki 2014.

In some analyses, the Hacrobia group (Haptophyta + Cryptophyta) is placed next to Archaeplastida, but in other ones it is nested inside the Archaeplastida. However, several recent studies have concluded that Haptophyta and Cryptophyta do not form a monophyletic group. The former could be a sister group to the SAR group, the latter cluster with the Archaeplastida (plants in the broad sense). As of February 2012, it remains unclear whether the Hacrobia forms a monophyletic group.

The division of the eukaryotes into two primary clades, bikonts (Archaeplastida + SAR + Excavata) and unikonts (Amoebozoa + Opisthokonta), derived from an ancestral biflagellar organism and an ancestral uniflagellar organism, respectively, had been suggested earlier. A 2012 study produced a somewhat similar division, although noting that the terms "unikonts" and "bikonts" were not used in the original sense.

Cavalier-Smith's Tree

Cavalier-Smith 2010, 2013, and 2014 places the eukaryotic tree's root between Excavata (with ventral feeding groove supported by a microtubular root) and the grooveless Euglenozoa:

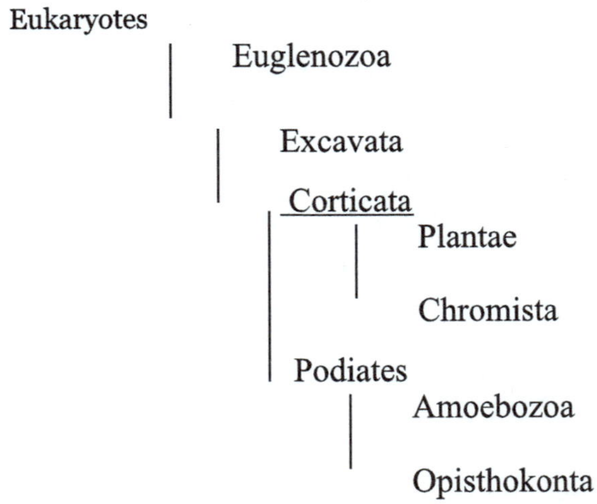

Alternative Views

Other analyses place the SAR supergroup within an expanded Chromalveolata, although they differ on the placement of the resulting five groups. Rogozin *et al.* in 2009 produced the tree shown below, where the primary division is between the Archaeplastida and all other eukaryotes.

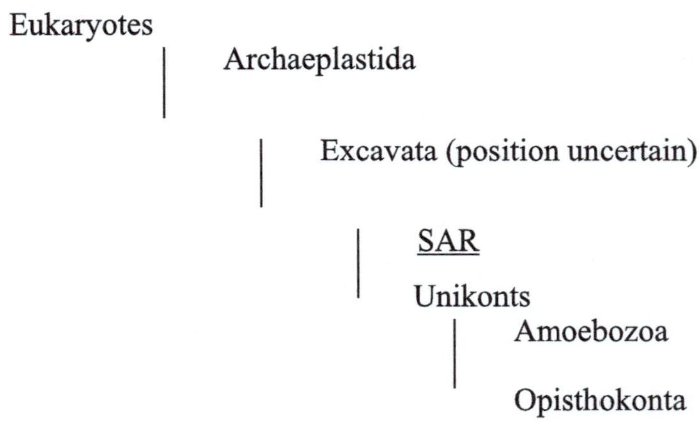

A paper published in 2009, which re-examined the data used in some of the analyses presented above as well as performing new ones, strongly suggested that the Archaeplastida are polyphyletic. The phylogeny finally proposed in the paper is shown below.

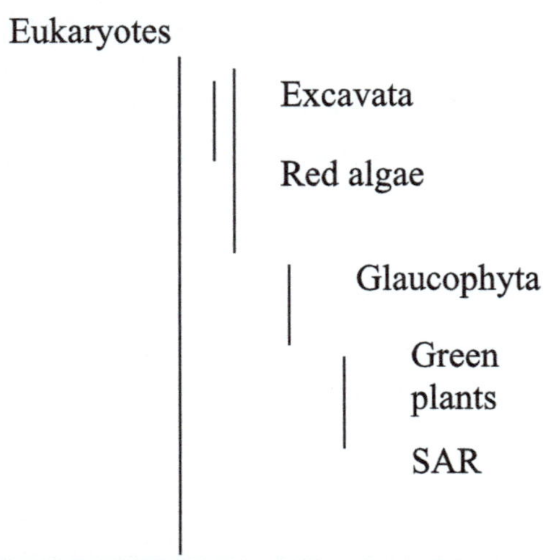

There are also the cladistic analyses of Diana Lipscomb based on classical data which have red algae as basal and Archeoplastida as paraphyletic. In the survey by Parfrey et al. it is recovered in only 42.6% of the molecular analyses that include it, that is, 26 out of 61. It is not recovered in Goloboff et al.'s combined analysis, and is mostly weakly supported in other molecular analyses. Laura Parfrey et al. point out that Archeoplastida support comes primarily from phylogenomic analyses and these may be picking up misleading endosymbiotic gene transfer signal of genes independently transferred from the plastid to the host nucleus in the 3 archeoplastid clades. And Stiller and Harrell emphasize that the group can be explained by "short-branch exclusion" and "subtle and easily overlooked biases can dominate the overall results of molecular phylogenetic analyses of ancient eukaryotic relationships. Sources of potential phylogenetic artifact should be investigated routinely, not just when obvious 'long-branch attraction' is encountered."

Red algae as basal is also supported by molecular evidence. *Cyanidioschyzon*, a red alga, is considered basal (sister group to the rest of eukaryotes) by Nagashima et al. and Seckbach. It has the most primitive chloroplast, only 1 mitochondrion, no vacuoles, no trienoic acids, and the smallest eukaryotic genome at 8 Mbp.

Rhizaria are only moderately supported (65.5% of the studies, i.e., 19 of 29), statistical support for them is inconsistent in multigene genealogies with larger taxon sampling and the group is ambiguously supported in Goloboff et al.

Origin of Eukaryotes

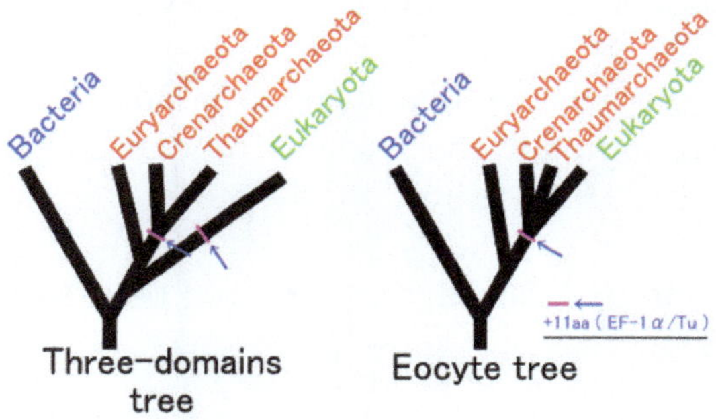

The three-domains tree and the Eocyte hypothesis.

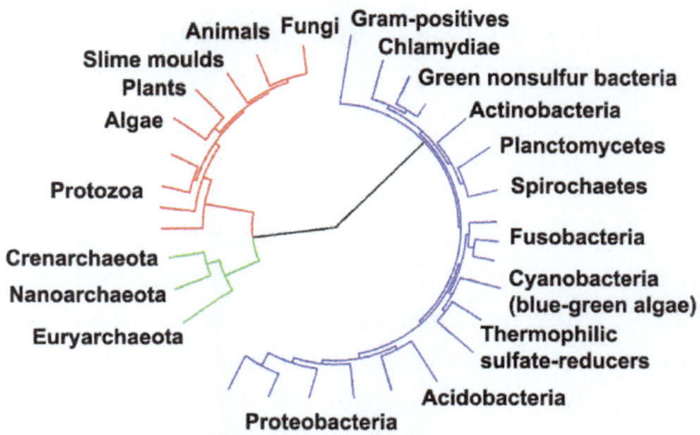

Phylogenetic tree showing the relationship between the eukaryotes and other forms of life. Eukaryotes are colored red, archaea green and bacteria blue.

Fossils

The origin of the eukaryotic cell is considered a milestone in the evolution of life, since eukaryotes include all complex cells and almost all multicellular organisms. The timing of this series of events is hard to determine; Knoll (2006) suggests they developed approximately 1.6–2.1 billion years ago. Some acritarchs are known from at least 1.65 billion years ago, and the possible alga *Grypania* has been found as far back as 2.1 billion years ago. The *Geosiphon*-like fossil fungus *Diskagma* has been found in paleosols 2.2 billion years old

Organized living structures have been found in the black shales of the Palaeoproterozoic Francevillian B Formation in Gabon, dated at 2.1 billion years old. Eukaryotic life could have evolved at that time. Fossils that are clearly related to modern groups start appearing an estimated 1.2 billion years ago, in the form of a red alga, though recent work suggests the existence of fossilized filamentous algae in the Vindhya basin dating back perhaps to 1.6 to 1.7 billion years ago.

Biomarkers suggest that at least stem eukaryotes arose even earlier. The presence of steranes in Australian shales indicates that eukaryotes were present in these rocks dated at 2.7 billion years old.

Relationship to Archaea

Eukaryotes are more closely related to Archaea than Bacteria, at least in terms of nuclear DNA and genetic machinery, and one controversial idea is to place them with Archaea in the clade Neomura. However, in other respects, such as membrane composition, eukaryotes are similar to Bacteria. Three main explanations for this have been proposed:

- Eukaryotes resulted from the complete fusion of two or more cells, wherein the cytoplasm formed from a eubacterium, and the nucleus from an archaeon, from a virus, or from a pre-cell.
- Eukaryotes developed from Archaea, and acquired their eubacterial characteristics from the proto-mitochondrion.
- Eukaryotes and Archaea developed separately from a modified eubacterium.

The chronocyte hypothesis for the origin of the eukaryotic cell postulates that a primitive eukaryotic cell was formed by the endosymbiosis of both archaea and bacteria by a third type of cell, termed a chronocyte.

Endomembrane System and Mitochondria

The origins of the endomembrane system and mitochondria are also unclear. The phagotrophic hypothesis proposes that eukaryotic-type membranes lacking a cell wall originated first, with the development of endocytosis, whereas mitochondria were acquired by ingestion as endosymbionts. The syntrophic hypothesis proposes that the proto-eukaryote relied on the proto-mitochondrion for food, and so ultimately grew to surround it. Here the membranes originated after the engulfment of the mitochondrion, in part thanks to mitochondrial genes (the hydrogen hypothesis is one particular version).

In a study using genomes to construct supertrees, Pisani *et al.* (2007) suggest that, along with evidence that there was never a mitochondrion-less eukaryote, eukaryotes evolved from a syntrophy between an archaea closely related to Thermoplasmatales and an α-proteobacterium, likely a symbiosis driven by sulfur or hydrogen. The mitochondrion and its genome is a remnant of the α-proteobacterial endosymbiont.

Hypotheses for the origin of Eukaryotes

Different hypotheses have been proposed as to how eukaryotic cells came into existence. These hypotheses can be classified into two distinct classes – autogenous models and chimeric models.

Autogenous Models

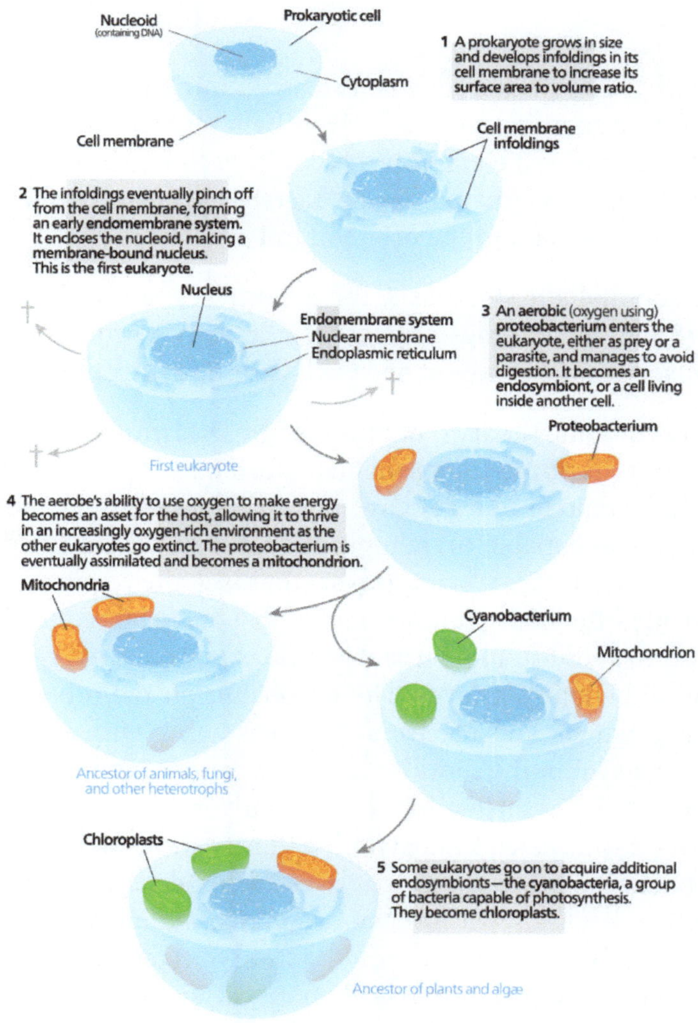

An autogenous model for the origin of eukaryotes.

Autogenous models propose that a proto-eukaryotic cell containing a nucleus existed first, and later acquired mitochondria. According to this model, a large prokaryote developed invaginations in its plasma membrane in order to obtain enough surface area to service its cytoplasmic volume.

As the invaginations differentiated in function, some became separate compartments—giving rise to the endomembrane system, including the endoplasmic reticulum, golgi apparatus, nuclear membrane, and single membrane structures such as lysosomes. Mitochondria are proposed to come from the endosymbiosis of an aerobic proteobacterium, and it's assumed that all the eukaryotic lineages that did not acquire mitochondria became extinct. Chloroplasts came about from another endosymbiotic event involving cyanobacteria. Since all eukaryotes have mitochondria, but not all have chloroplasts, mitochondria are thought to have come first. This is the serial endosymbiosis theory.

Some models propose that the origins of double layered organelles, such as mitochondria and chloroplasts, in the proto-eukaryotic cell is due to the compartmentalization of DNA vesicles that were formed from the secondary invaginations or more detailed infoldings of cellular membrane.

Chimeric Models

Chimeric models claim that two prokaryotic cells existed initially – an archaeon and a bacterium. These cells underwent a merging process, either by a physical fusion or by endosymbiosis, thereby leading to the formation of a eukaryotic cell. Within these chimeric models, some studies further claim that mitochondria originated from a bacterial ancestor while others emphasize the role of endosymbiotic processes behind the origin of mitochondria.

Based on the process of mutualistic symbiosis, the hypotheses can be categorized as – the serial endosymbiotic theory (SET), the hydrogen hypothesis (mostly a process of symbiosis where hydrogen transfer takes place among different species), and the syntrophy hypothesis.

According to serial endosymbiotic theory (championed by Dr. Lynn Margulis), a union between a motile anaerobic bacterium (like *Spirochaeta*) and a thermoacidophilic crenarchaeon (like *Thermoplasma* which is sulfidogenic in nature) gave rise to the present day eukaryotes. This union established a motile organism capable of living in the already existing acidic and sulfurous waters. Oxygen is known to cause toxicity to organisms that lack the required metabolic machinery. Thus, the archaeon provided the bacterium with a highly beneficial reduced environment (sulfur and sulfate were reduced to sulfide). In microaerophilic conditions, oxygen was reduced to water thereby creating a mutual benefit platform. The bacterium on the other hand, contributed the necessary fermentation products and electron acceptors along with its motility feature to the archaeon thereby gaining a swimming motility for the organism. From a consortium of bacterial and archaeal DNA originated the nuclear genome of eukaryotic cells. Spirochetes gave rise to the motile features of eukaryotic cells. Endosymbiotic unifications of the ancestors of alpha-proteobacteria and cyanobacteria, led to the origin of mitochondria and plastids respectively. For example, *Thiodendron* has been known to have originated via an ectosymbiotic process based on a similar syntrophy of sulfur existing between the two types of bacteria – *Desulphobacter* and *Spirochaeta*. However, such an association based on motile symbiosis have never been observed practically. Also there is no evidence of archaeans and spirochetes adapting to intense acid-based environments.

In the hydrogen hypothesis, the symbiotic linkage of an anaerobic and autotrophic methanogenic archaeon (host) with an alpha-proteobacterium (the symbiont) gave rise to the eukaryotes. The host utilized hydrogen (H_2) and carbon dioxide (CO_2) to produce methane while the symbiont, capable of aerobic respiration, expelled H_2 and CO_2 as byproducts of anaerobic fermentation process. The host's methanogenic environment worked as a sink for H_2, which resulted in heightened bacterial fermentation. Endosymbiotic gene transfer (EGT) acted as a catalyst for the host to acquire the symbionts' carbohydrate metabolism and turn heterotrophic in nature. Subsequently, the host's methane forming capability was lost. Thus, the origins of the heterotrophic organelle (symbiont) are identical to the origins of the eukaryotic lineage. In this hypothesis, the presence of H_2 represents the selective force that forged eukaryotes out of prokaryotes.

The syntrophy hypothesis was developed in contrast to the hydrogen hypothesis and proposes the existence of two symbiotic events. According to this theory, eukaryogenesis (i.e. origin of eukaryotic cells) occurred based on metabolic symbiosis (syntrophy) between a methanogenic archaeon and a delta-proteobacterium. This syntrophic symbiosis was initially facilitated by H_2 transfer between different species under anaerobic environments. In earlier stages, an alpha-proteobacterium became a member of this integration, and later developed into the mitochondrion. Gene transfer from a delta-proteobacterium to an archaeon led to the methanogenic archaeon developing into a nucleus. The archaeon constituted the genetic apparatus while the delta-proteobacterium contributed towards the cytoplasmic features. This theory incorporates two selective forces that were needed to be considered during the time of nucleus evolution – (a) presence of metabolic partitioning in order to avoid the harmful effects of the co-existence of anabolic and catabolic cellular pathways, and (b) prevention of abnormal biosynthesis of proteins that occur due to a vast spread of introns in the archaeal genes after acquiring the mitochondrion and the loss of methanogenesis.

Thus, the origin of eukaryotes by endosymbiotic processes has been broadly recognized and accepted so far. Mitochondria and plastids have been known to originate from a bacterial ancestor during parallel adaptation to anaerobiosis. However, there still remains a greater need in assessing the question of how much eukaryotic complexity is being originated via an implementation of these symbiogenetic theories.

References

- Bauman, Robert W.; Tizard, Ian R.; Machunis-Masouka, Elizabeth (2006). Microbiology. San Francisco: Pearson Benjamin Cummings. ISBN 0-8053-7693-3.
- Madigan, Michael T. (2012). Brock biology of microorganisms (13th ed.). San Francisco: Benjamin Cummings. ISBN 9780321649638.
- "Direct Observations". The Biofilm Primer. Springer Series on Biofilms. 1. 2007. pp. 3–4. doi:10.1007/978-3-540-68022-2_2. ISBN 978-3-540-68021-5.
- Nelson, David L.; Cox, Michael M. (2005). Lehninger Principles of Biochemistry (4th ed.). New York: W.H. Freeman. ISBN 0-7167-4339-6.
- Martin, E.A., ed. (1983). Macmillan Dictionary of Life Sciences (2nd ed.). London: Macmillan Press. ISBN 0-333-34867-2.
- Lynn Margulis, Heather I. McKhann & Lorraine Olendzenski (ed.), Illustrated Glossary of Protoctista, Jones and Bartlett Publishers, Boston, 1993, p.xviii. ISBN 0-86720-081-2
- Sanders, Robert (11 April 2016). "Wealth of unsuspected new microbes expands tree of life". Berkeley News. Retrieved 11 April 2016.
- Hug, Laura A.; Baker, Brett J.; Anantharaman, Karthik; Brown, Christopher T.; Probst, Alexander J.; Castelle,

Cindy J.; Butterfield, Cristina N.; Hernsdorf, Alex W.; Amano, Yuki; Ise, Kotaro; Suzuki, Yohey; Dudek, Natasha; Relman, David A.; Finstad, Kari M.; Amundson, Ronald; Thomas, Brian C.; Banfield, Jillian F. (11 April 2016). "A new view of the tree of life". Nature (16048). doi:10.1038/nmicrobiol.2016.48.

- Davis, Josh L. (13 May 2016). "Scientists Shocked To Discover Eukaryote With NO Mitochondria". IFL Science. Retrieved 13 May 2016.
- Burki, F. (2014). "The eukaryotic tree of life from a global phylogenomic perspective". Cold Spring Harbor Perspectives in Biology. 6: 1–17. doi:10.1101/cshperspect.a016147.
- S, J Ayala (April 1, 1994). "Transport and internal organization of membranes: vesicles, membrane networks and GTP-binding proteins". Journal of Cell Science. 107 (107): 753–763. PMID 8056835. Retrieved 27 March 2013. .
- Jagus, R; Bachvaroff, TR; Joshi, B; Place, AR (2012). "Diversity of eukaryotic translational initiation factor eIF4E in protists". Comp Funct Genomics. 2012: 134839. doi:10.1155/2012/134839.
- Latorre, A.; Durban, A.; Moya, A.; Pereto, J. (2011). "The role of symbiosis in eukaryotic evolution. Origins and evolution of life – An astrobiological perspective": 326–339

3

Fundamentals of Cell Structure

An evolved cell has a particular structure that houses components such as the cell nucleus, golgi apparatus and the cytoskeleton. All these contribute to the functioning of the cell. The topics discussed in this chapter are of great importance to broaden the existing knowledge on cell structure.

Organelle

In cell biology, an organelle is a specialized subunit within a cell that has a specific function. Individual organelles are usually separately enclosed within their own lipid bilayers.

The name *organelle* comes from the idea that these structures are to cells what an organ is to the body (hence the name *organelle,* the suffix *-elle* being a diminutive). Organelles are identified by microscopy, and can also be purified by cell fractionation. There are many types of organelles, particularly in eukaryotic cells. While prokaryotes do not possess organelles *per se*, some do contain protein-based microcompartments, which are thought to act as primitive organelles.

History and Terminology

In biology *organs* are defined as confined functional units within an organism. The analogy of bodily organs to microscopic cellular substructures is obvious, as from even early works, authors of respective textbooks rarely elaborate on the distinction between the two.

Credited as the first to use a diminutive of *organ* (i.e., little organ) for cellular structures was German zoologist Karl August Möbius (1884), who used the term *organula* (plural of *organulum*, the diminutive of Latin *organum*). In a footnote, which was published as a correction in the next issue of the journal, he justified his suggestion to call organs of unicellular organisms "organella" since they are only differently formed parts of one cell, in contrast to multicellular organs of multicellular organisms.

Types of Organelles

While most cell biologists consider the term organelle to be synonymous with "cell compartment", other cell biologists choose to limit the term organelle to include only those that are DNA-containing, having originated from formerly autonomous microscopic organisms acquired via endosymbiosis.

Under this definition, there would only be two broad classes of organelles (i.e. those that contain their own DNA, and have originated from endosymbiotic bacteria):

- mitochondria (in almost all eukaryotes)
- plastids (e.g. in plants, algae, and some protists).

Other organelles are also suggested to have endosymbiotic origins, but do not contain their own DNA.

Under the more restricted definition of membrane-bound structures, some parts of the cell do not qualify as organelles. Nevertheless, the use of organelle to refer to non-membrane bound structures such as ribosomes is common. This has led some texts to delineate between membrane-bound and non-membrane bound organelles. The non-membrane bound organelles, also called large biomolecular complexes, are large assemblies of macromolecules that carry out particular and specialized functions, but they lack membrane boundaries. Such cell structures include:

- large RNA and protein complexes: ribosome, spliceosome, vault
- large protein complexes: proteasome, DNA polymerase III holoenzyme, RNA polymerase II holoenzyme, symmetric viral capsids, complex of GroEL and GroES; membrane protein complexes: photosystem I, ATP synthase
- large DNA and protein complexes: nucleosome
- centriole and microtubule-organizing center (MTOC)
- cytoskeleton
- flagellum
- Nucleolus
- Stress granule
- Germ cell granule
- Neuronal transport granule

Eukaryotic Organelles

Eukaryotic cells are structurally complex, and by definition are organized, in part, by interior compartments that are themselves enclosed by lipid membranes that resemble the outermost cell membrane. The larger organelles, such as the nucleus and vacuoles, are easily visible with the light microscope. They were among the first biological discoveries made after the invention of the microscope.

Not all eukaryotic cells have each of the organelles listed below. Exceptional organisms have cells that do not include some organelles that might otherwise be considered universal to eukaryotes (such as mitochondria). There are also occasional exceptions to the number of membranes surrounding organelles, listed in the tables below (e.g., some that are listed as double-membrane are sometimes found with single or triple membranes). In addition, the number of individual organelles of each type found in a given cell varies depending upon the function of that cell.

Major eukaryotic organelles

Organelle	Main function	Structure	Organisms	Notes
chloroplast (plastid)	photosynthesis, traps energy from sunlight	double-membrane compartment	plants, protists (rare kleptoplastic organisms)	has own DNA; theorized to be engulfed by the ancestral eukaryotic cell (endosymbiosis)
endoplasmic reticulum	translation and folding of new proteins (rough endoplasmic reticulum), expression of lipids (smooth endoplasmic reticulum)	single-membrane compartment	all eukaryotes	rough endoplasmic reticulum is covered with ribosomes, has folds that are flat sacs; smooth endoplasmic reticulum has folds that are tubular
Flagellum	locomotion, sensory		some eukaryotes	
Golgi apparatus	sorting, packaging, processing and modification of proteins	single-membrane compartment	all eukaryotes	cis-face (convex) nearest to rough endoplasmic reticulum; trans-face (concave) farthest from rough endoplasmic reticulum
mitochondria	energy production from the oxidation of glucose substances and the release of adenosine triphosphate	double-membrane compartment	most eukaryotes	has own DNA; theorized to be engulfed by an ancestral eukaryotic cell (endosymbiosis)
vacuole	storage, transportation, helps maintain homeostasis	single-membrane compartment	eukaryotes	
nucleus	DNA maintenance, controls all activities of the cell, RNA transcription	double-membrane compartment	all eukaryotes	

contains bulk of genome Mitochondria and chloroplasts, which have double-membranes and their own DNA, are believed to have originated from incompletely consumed or invading prokaryotic organisms, which were adopted as a part of the invaded cell. This idea is supported in the Endosymbiotic theory.

Minor eukaryotic organelles and cell components

Organelle/Macromolecule	Main function	Structure	Organisms
acrosome	helps spermatozoa fuse with ovum	single-membrane compartment	many animals
autophagosome	vesicle that sequesters cytoplasmic material and organelles for degradation	double-membrane compartment	all eukaryotes
centriole	anchor for cytoskeleton, organizes cell division by forming spindle fibers	Microtubule protein	animals
cilium	movement in or of external medium; "critical developmental signaling pathway".	Microtubule protein	animals, protists, few plants

eyespot apparatus	detects light, allowing phototaxis to take place		green algae and other unicellular photosynthetic organisms such as euglenids
glycosome	carries out glycolysis	single-membrane compartment	Some protozoa, such as *Trypanosomes*.
glyoxysome	conversion of fat into sugars	single-membrane compartment	plants
hydrogenosome	energy & hydrogen production	double-membrane compartment	a few unicellular eukaryotes
lysosome	breakdown of large molecules (e.g., proteins + polysaccharides)	single-membrane compartment	most eukaryotes
melanosome	pigment storage	single-membrane compartment	animals
mitosome	probably plays a role in Iron-sulfur cluster (Fe-S) assembly	double-membrane compartment	a few unicellular eukaryotes that lack mitochondria
myofibril	myocyte contraction	bundled filaments	animals
nematocyst	stinging	coiled hollow tubule	Cnidarians
nucleolus	pre-ribosome production	protein-DNA-RNA	most eukaryotes
parenthesome	not characterized	not characterized	fungi
peroxisome	breakdown of metabolic hydrogen peroxide	single-membrane compartment	all eukaryotes
proteasome	degradation of unneeded or damaged proteins by proteolysis	very large protein complex	All eukaryotes, all archaea, some bacteria
ribosome (80S)	translation of RNA into proteins	RNA-protein	all eukaryotes
vesicle	material transport	single-membrane compartment	all eukaryotes
Stress granule	mRNA storage	membraneless (mRNP complexes)	Most eukaryotes

Other Related Structures:

- Cytosol
- Endomembrane System
- Nucleosome
- Microtubule
- Cell Membrane

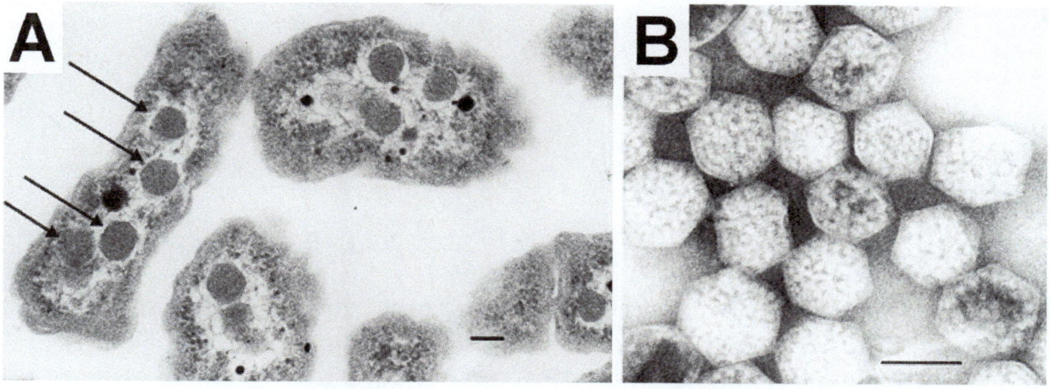

(A) Electron Micrograph Of *Halothiobacillus Neapolitanus* Cells, Arrows Highlight Carboxysomes. (B) Image Of Intact Carboxysomes Isolated From *H. Neapolitanus*. Scale Bars Are 100 Nm.

Prokaryotic Organelles

Prokaryotes are not as structurally complex as eukaryotes, and were once thought not to have any internal structures enclosed by lipid membranes. In the past, they were often viewed as having little internal organization, but slowly, details are emerging about prokaryotic internal structures. An early false turn was the idea developed in the 1970s that bacteria might contain membrane folds termed mesosomes, but these were later shown to be artifacts produced by the chemicals used to prepare the cells for electron microscopy.

However, more recent research has revealed that at least some prokaryotes have microcompartments such as carboxysomes. These subcellular compartments are 100–200 nm in diameter and are enclosed by a shell of proteins. Even more striking is the description of membrane-bound magnetosomes in bacteria, reported in 2006, as well as the nucleus-like structures of the *Planctomycetes* that are surrounded by lipid membranes, reported in 2005.

Prokaryotic Organelles and cell Components

Organelle/ Macromolecule	Main function	Structure	Organisms
carboxysome	carbon fixation	protein-shell compartment	some bacteria
chlorosome	photosynthesis	light harvesting complex	green sulfur bacteria
flagellum	movement in external medium	protein filament	some prokaryotes and eukaryotes
magnetosome	magnetic orientation	inorganic crystal, lipid membrane	magnetotactic bacteria
nucleoid	DNA maintenance, transcription to RNA	DNA-protein	prokaryotes
plasmid	DNA exchange	circular DNA	some bacteria
ribosome (70S)	translation of RNA into proteins	RNA-protein	bacteria and archaea
thylakoid	photosynthesis	photosystem proteins and pigments	mostly cyanobacteria

mesosomes	functions of Golgi bodies, centrioles, etc.	small irregular shaped organelle containing ribosomes	present in most prokaryotic cells
Pilus	Adhesion to other cells for conjugation or to a solid substrate to create motile forces.	a hair-like appendage sticking out (though partially embedded into) the plasma membrane	prokaryotic cells

Proteins and Organelles

The function of a protein is closely correlated with the organelle in which it resides. Some methods were proposed for predicting the organelle in which an uncharacterized protein is located according to its amino acid composition and some methods were based on pseudo amino acid composition.

Cell Nucleus

In cell biology, the nucleus (pl. *nuclei*; from Latin *nucleus* or *nuculeus*, meaning kernel) is a membrane-enclosed organelle found in eukaryotic cells. Eukaryotes usually have a single nucleus, but a few cell types, such as mammalian red blood cells, have no nuclei, and a few others have many.

Cell nuclei contain most of the cell's genetic material, organized as multiple long linear DNA molecules in complex with a large variety of proteins, such as histones, to form chromosomes. The genes within these chromosomes are the cell's nuclear genome and are structured in such a way to promote cell function. The nucleus maintains the integrity of genes and controls the activities of the cell by regulating gene expression—the nucleus is, therefore, the control center of the cell. The main structures making up the nucleus are the nuclear envelope, a double membrane that encloses the entire organelle and isolates its contents from the cellular cytoplasm, and the nuclear matrix (which includes the nuclear lamina), a network within the nucleus that adds mechanical support, much like the cytoskeleton, which supports the cell as a whole.

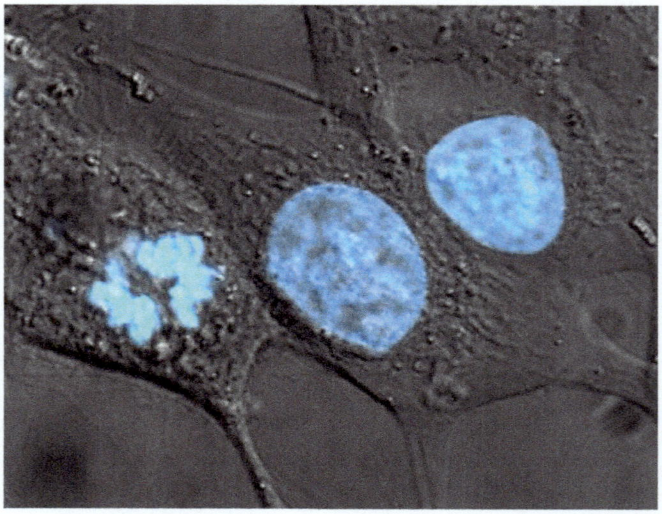

HeLa cells stained for nuclear DNA with the Blue Hoechst dye. The central and rightmost cell are in interphase, thus their entire nuclei are labeled. On the left, a cell is going through mitosis and its DNA has condensed.

Because the nuclear membrane is impermeable to large molecules, nuclear pores are required that regulate nuclear transport of molecules across the envelope. The pores cross both nuclear membranes, providing a channel through which larger molecules must be actively transported by carrier proteins while allowing free movement of small molecules and ions. Movement of large molecules such as proteins and RNA through the pores is required for both gene expression and the maintenance of chromosomes. Although the interior of the nucleus does not contain any membrane-bound sub compartments, its contents are not uniform, and a number of *sub-nuclear bodies* exist, made up of unique proteins, RNA molecules, and particular parts of the chromosomes. The best-known of these is the nucleolus, which is mainly involved in the assembly of ribosomes. After being produced in the nucleolus, ribosomes are exported to the cytoplasm where they translate mRNA.

History

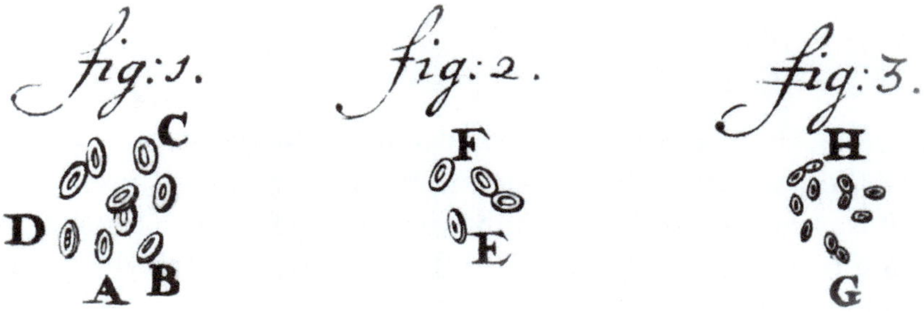

Oldest known depiction of cells and their nuclei by Antonie van Leeuwenhoek, 1719

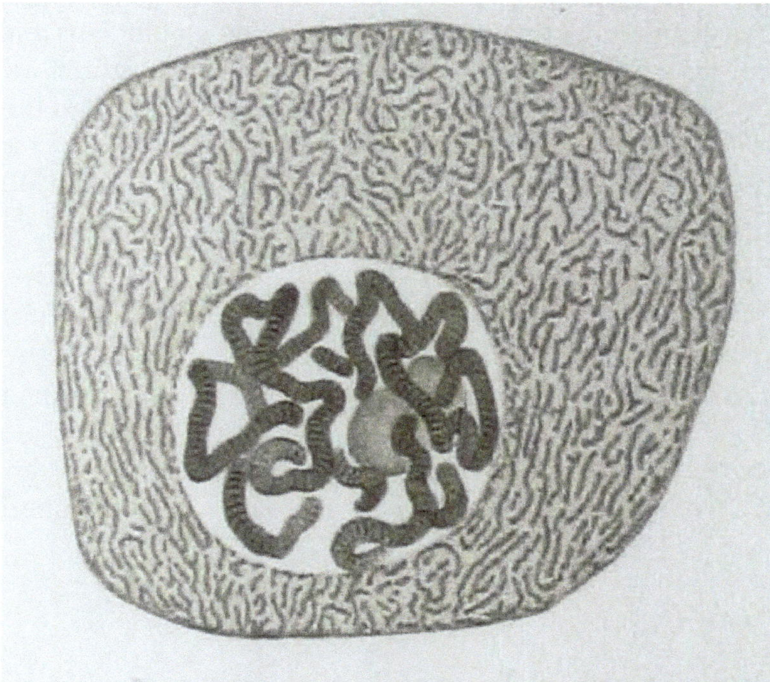

Drawing of a *Chironomus* salivary gland cell published by Walther Flemming in 1882. The nucleus contains Polytene chromosomes.

The nucleus was the first organelle to be discovered. What is most likely the oldest preserved drawing dates back to the early microscopist Antonie van Leeuwenhoek (1632–1723). He observed a "lumen", the nucleus, in the red blood cells of salmon. Unlike mammalian red blood cells, those of other vertebrates still contain nuclei.

The nucleus was also described by Franz Bauer in 1804 and in more detail in 1831 by Scottish botanist Robert Brown in a talk at the Linnean Society of London. Brown was studying orchids under microscope when he observed an opaque area, which he called the "areola" or "nucleus", in the cells of the flower's outer layer.

He did not suggest a potential function. In 1838, Matthias Schleiden proposed that the nucleus plays a role in generating cells, thus he introduced the name "cytoblast" (cell builder). He believed that he had observed new cells assembling around "cytoblasts". Franz Meyen was a strong opponent of this view, having already described cells multiplying by division and believing that many cells would have no nuclei. The idea that cells can be generated de novo, by the "cytoblast" or otherwise, contradicted work by Robert Remak (1852) and Rudolf Virchow (1855) who decisively propagated the new paradigm that cells are generated solely by cells ("Omnis cellula e cellula"). The function of the nucleus remained unclear.

Between 1877 and 1878, Oscar Hertwig published several studies on the fertilization of sea urchin eggs, showing that the nucleus of the sperm enters the oocyte and fuses with its nucleus. This was the first time it was suggested that an individual develops from a (single) nucleated cell. This was in contradiction to Ernst Haeckel's theory that the complete phylogeny of a species would be repeated during embryonic development, including generation of the first nucleated cell from a "monerula", a structureless mass of primordial mucus ("Urschleim"). Therefore, the necessity of the sperm nucleus for fertilization was discussed for quite some time. However, Hertwig confirmed his observation in other animal groups, including amphibians and molluscs. Eduard Strasburger produced the same results for plants in 1884. This paved the way to assign the nucleus an important role in heredity. In 1873, August Weismann postulated the equivalence of the maternal and paternal germ *cells* for heredity. The function of the nucleus as carrier of genetic information became clear only later, after mitosis was discovered and the Mendelian rules were rediscovered at the beginning of the 20th century; the chromosome theory of heredity was therefore developed.

Structures

The nucleus is the largest cellular organelle in animal cells. In mammalian cells, the average diameter of the nucleus is approximately 6 micrometres (μm), which occupies about 10% of the total cell volume. The viscous liquid within it is called nucleoplasm, and is similar in composition to the cytosol found outside the nucleus. It appears as a dense, roughly spherical or irregular organelle.

Nuclear Envelope and Pores

The nuclear envelope, otherwise known as nuclear membrane, consists of two cellular membranes, an inner and an outer membrane, arranged parallel to one another and separated by 10 to 50 nanometres (nm). The nuclear envelope completely encloses the nucleus and separates the cell's genetic material from the surrounding cytoplasm, serving as a barrier to prevent macromolecules from diffusing freely between the nucleoplasm and the cytoplasm. The outer nuclear membrane is continuous with the membrane of the rough endoplasmic reticulum (RER), and is similarly studded with ribosomes. The space between the membranes is called the perinuclear space and is continuous with the RER lumen.

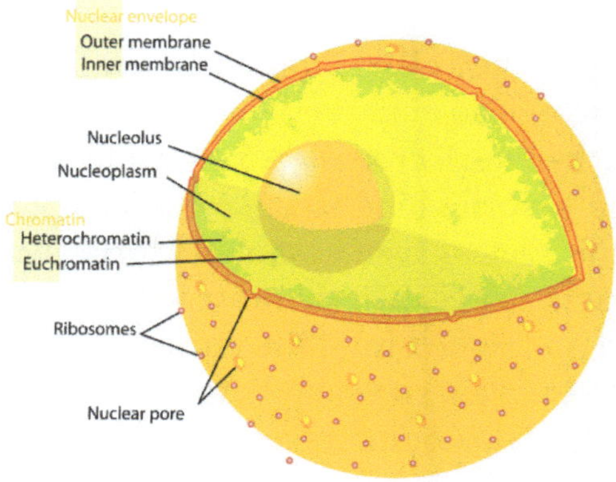

The eukaryotic cell nucleus. Visible in this diagram are the ribosome-studded double membranes of the nuclear envelope, the DNA (complexed as chromatin), and the nucleolus. Within the cell nucleus is a viscous liquid called nucleoplasm, similar to the cytoplasm found outside the nucleus.

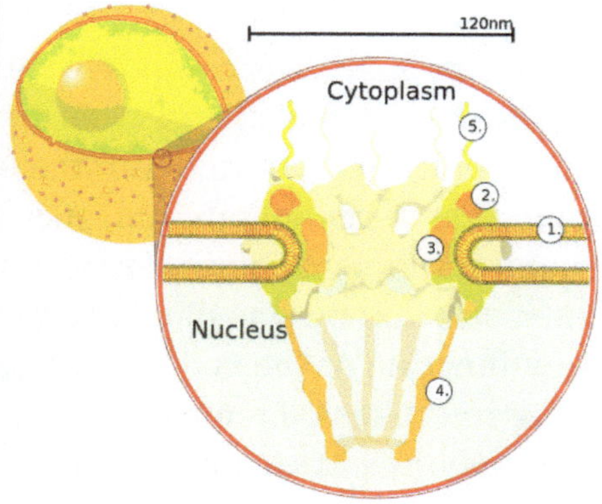

A cross section of a nuclear pore on the surface of the nuclear envelope (1). Other diagram labels show (2) the outer ring, (3) spokes, (4) basket, and (5) filaments.

Nuclear pores, which provide aqueous channels through the envelope, are composed of multiple proteins, collectively referred to as nucleoporins. The pores are about 125 million daltons in molecular weight and consist of around 50 (in yeast) to several hundred proteins (in vertebrates). The pores are 100 nm in total diameter; however, the gap through which molecules freely diffuse is only about 9 nm wide, due to the presence of regulatory systems within the center of the pore. This size selectively allows the passage of small water-soluble molecules while preventing larger molecules, such as nucleic acids and larger proteins, from inappropriately entering or exiting the nucleus. These large molecules must be actively transported into the nucleus instead. The nucleus of a typical mammalian cell will have about 3000 to 4000 pores throughout its envelope, each of which contains an eightfold-symmetric ring-shaped structure at a position where the inner and outer membranes fuse. Attached to the ring is a structure called the *nuclear basket* that extends into the nucleoplasm, and a series of filamentous extensions that reach into the cytoplasm. Both structures serve to mediate binding to nuclear transport proteins.

Most proteins, ribosomal subunits, and some DNAs are transported through the pore complexes in a process mediated by a family of transport factors known as karyopherins. Those karyopherins that mediate movement into the nucleus are also called importins, whereas those that mediate movement out of the nucleus are called exportins. Most karyopherins interact directly with their cargo, although some use adaptor proteins. Steroid hormones such as cortisol and aldosterone, as well as other small lipid-soluble molecules involved in intercellular signaling, can diffuse through the cell membrane and into the cytoplasm, where they bind nuclear receptor proteins that are trafficked into the nucleus. There they serve as transcription factors when bound to their ligand; in the absence of ligand, many such receptors function as histone deacetylases that repress gene expression.

Nuclear Lamina

In animal cells, two networks of intermediate filaments provide the nucleus with mechanical support: The nuclear lamina forms an organized meshwork on the internal face of the envelope, while less organized support is provided on the cytosolic face of the envelope. Both systems provide structural support for the nuclear envelope and anchoring sites for chromosomes and nuclear pores.

The nuclear lamina is composed mostly of lamin proteins. Like all proteins, lamins are synthesized in the cytoplasm and later transported to the nucleus interior, where they are assembled before being incorporated into the existing network of nuclear lamina. Lamins found on the cytosolic face of the membrane, such as emerin and nesprin, bind to the cytoskeleton to provide structural support. Lamins are also found inside the nucleoplasm where they form another regular structure, known as the *nucleoplasmic veil*, that is visible using fluorescence microscopy. The actual function of the veil is not clear, although it is excluded from the nucleolus and is present during interphase. Lamin structures that make up the veil, such as LEM3, bind chromatin and disrupting their structure inhibits transcription of protein-coding genes.

Like the components of other intermediate filaments, the lamin monomer contains an alpha-helical domain used by two monomers to coil around each other, forming a dimer structure called a coiled coil. Two of these dimer structures then join side by side, in an antiparallel arrangement, to form a tetramer called a *protofilament*. Eight of these protofilaments form a lateral arrangement that is twisted to form a ropelike *filament*. These filaments can be assembled or disassembled in a dynamic manner, meaning that changes in the length of the filament depend on the competing rates of filament addition and removal.

Mutations in lamin genes leading to defects in filament assembly cause a group of rare genetic disorders known as *laminopathies*. The most notable laminopathy is the family of diseases known as progeria, which causes the appearance of premature aging in its sufferers. The exact mechanism by which the associated biochemical changes give rise to the aged phenotype is not well understood.

Chromosomes

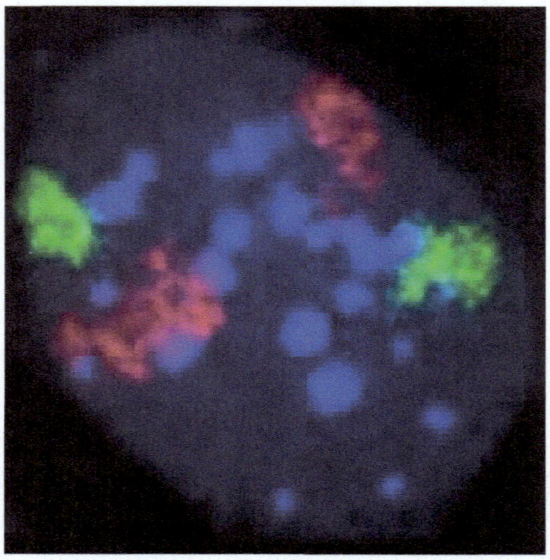

A mouse fibroblast nucleus in which DNA is stained blue. The distinct chromosome territories of chromosome 2 (red) and chromosome 9 (green) are stained with fluorescent in situ hybridization.

The cell nucleus contains the majority of the cell's genetic material in the form of multiple linear DNA molecules organized into structures called chromosomes. Each human cell contains roughly two meters of DNA. During most of the cell cycle these are organized in a DNA-protein complex known as chromatin, and during cell division the chromatin can be seen to form the well-defined chromosomes familiar from a karyotype. A small fraction of the cell's genes are located instead in the mitochondria.

There are two types of chromatin. Euchromatin is the less compact DNA form, and contains genes that are frequently expressed by the cell. The other type, heterochromatin, is the more compact form, and contains DNA that is infrequently transcribed. This structure is further categorized into *facultative* heterochromatin, consisting of genes that are organized as heterochromatin only in certain cell types or at certain stages of development, and *constitutive* heterochromatin that consists of chromosome structural components such as telomeres and centromeres. During interphase the chromatin organizes itself into discrete individual patches, called *chromosome territories*. Active genes, which are generally found in the euchromatic region of the chromosome, tend to be located towards the chromosome's territory boundary.

Antibodies to certain types of chromatin organization, in particular, nucleosomes, have been associated with a number of autoimmune diseases, such as systemic lupus erythematosus. These are known as anti-nuclear antibodies (ANA) and have also been observed in concert with multiple sclerosis as part of general immune system dysfunction. As in the case of progeria, the role played by the antibodies in inducing the symptoms of autoimmune diseases is not obvious.

Nucleolus

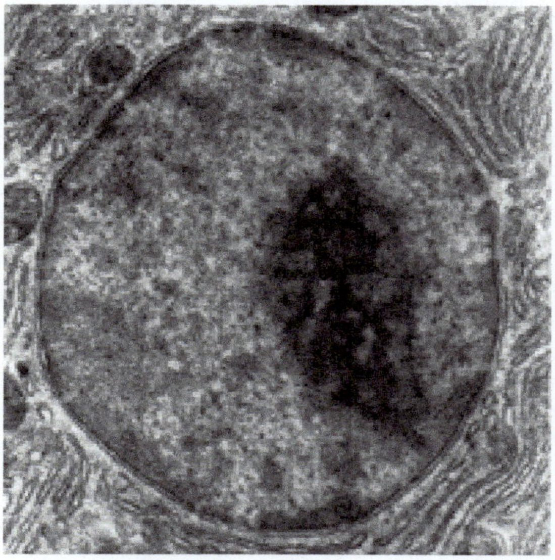

An electron micrograph of a cell nucleus, showing the darkly stained nucleolus

The nucleolus is a discrete densely stained structure found in the nucleus. It is not surrounded by a membrane, and is sometimes called a *suborganelle*. It forms around tandem repeats of rDNA, DNA coding for ribosomal RNA (rRNA). These regions are called nucleolar organizer regions (NOR). The main roles of the nucleolus are to synthesize rRNA and assemble ribosomes. The structural cohesion of the nucleolus depends on its activity, as ribosomal assembly in the nucleolus results in the transient association of nucleolar components, facilitating further ribosomal assembly, and hence further association. This model is supported by observations that inactivation of rDNA results in intermingling of nucleolar structures.

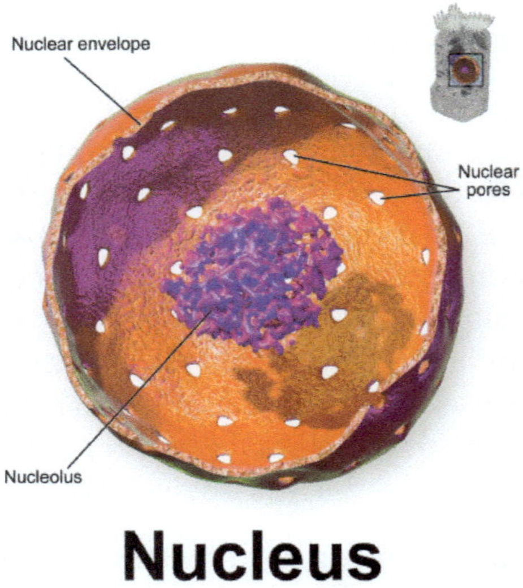

3D rendering of nucleus with location of nucleolus

In the first step of ribosome assembly, a protein called RNA polymerase I transcribes rDNA, which forms a large pre-rRNA precursor. This is cleaved into the subunits 5.8S, 18S, and 28S rRNA. The transcription, post-transcriptional processing, and assembly of rRNA occurs in the nucleolus, aided by small nucleolar RNA (snoRNA) molecules, some of which are derived from spliced introns from messenger RNAs encoding genes related to ribosomal function. The assembled ribosomal subunits are the largest structures passed through the nuclear pores.

When observed under the electron microscope, the nucleolus can be seen to consist of three distinguishable regions: the innermost *fibrillar centers* (FCs), surrounded by the *dense fibrillar component* (DFC), which in turn is bordered by the *granular component* (GC). Transcription of the rDNA occurs either in the FC or at the FC-DFC boundary, and, therefore, when rDNA transcription in the cell is increased, more FCs are detected. Most of the cleavage and modification of rRNAs occurs in the DFC, while the latter steps involving protein assembly onto the ribosomal subunits occur in the GC.

Other Subnuclear Bodies

Subnuclear structure sizes	
Structure name	**Structure diameter**
Cajal bodies	0.2–2.0 μm
PIKA	5 μm
PML bodies	0.2–1.0 μm
Paraspeckles	0.2–1.0 μm
Speckles	20–25 nm

Besides the nucleolus, the nucleus contains a number of other non-membrane-delineated bodies. These include Cajal bodies, Gemini of coiled bodies, polymorphic interphase karyosomal association (PIKA), promyelocytic leukaemia (PML) bodies, paraspeckles, and splicing speckles. Although little is known about a number of these domains, they are significant in that they show that the nucleoplasm is not a uniform mixture, but rather contains organized functional subdomains.

Other subnuclear structures appear as part of abnormal disease processes. For example, the presence of small intranuclear rods has been reported in some cases of nemaline myopathy. This condition typically results from mutations in actin, and the rods themselves consist of mutant actin as well as other cytoskeletal proteins.

Cajal Bodies and Gems

A nucleus typically contains between 1 and 10 compact structures called Cajal bodies or coiled bodies (CB), whose diameter measures between 0.2 μm and 2.0 μm depending on the cell type and species. When seen under an electron microscope, they resemble balls of tangled thread and are dense foci of distribution for the protein coilin. CBs are involved in a number of different roles relating to RNA processing, specifically small nucleolar RNA (snoRNA) and small nuclear RNA (snRNA) maturation, and histone mRNA modification.

Similar to Cajal bodies are Gemini of Cajal bodies, or gems, whose name is derived from the Gemini constellation in reference to their close "twin" relationship with CBs. Gems are similar in size and shape to CBs, and in fact are virtually indistinguishable under the microscope. Unlike CBs, gems do not contain small nuclear ribonucleoproteins (snRNPs), but do contain a protein called survival of motor neuron (SMN) whose function relates to snRNP biogenesis. Gems are believed to assist CBs in snRNP biogenesis, though it has also been suggested from microscopy evidence that CBs and gems are different manifestations of the same structure. Later ultrastructural studies have shown gems to be twins of Cajal bodies with the difference being in the coilin component; Cajal bodies are SMN positive and coilin positive, and gems are SMN positive and coilin negative.

RAFA and PTF Domains

RAFA domains, or polymorphic interphase karyosomal associations, were first described in microscopy studies in 1991. Their function remains unclear, though they were not thought to be associated with active DNA replication, transcription, or RNA processing. They have been found to often associate with discrete domains defined by dense localization of the transcription factor PTF, which promotes transcription of small nuclear RNA (snRNA).

PML Bodies

Promyelocytic leukaemia bodies (PML bodies) are spherical bodies found scattered throughout the nucleoplasm, measuring around 0.1–1.0 μm. They are known by a number of other names, including nuclear domain 10 (ND10), Kremer bodies, and PML oncogenic domains. PML bodies are named after one of their major components, the promyelocytic leukemia protein (PML). They are often seen in the nucleus in association with Cajal bodies and cleavage bodies. PML bodies belong to the nuclear matrix, an ill-defined super-structure of the nucleus proposed to anchor and regulate many nuclear functions, including DNA replication, transcription, or epigenetic silencing. The PML protein is the key organizer of these domains that recruits an ever-growing number of proteins, whose only common known feature to date is their ability to be SUMOylated. Yet, pml-/- mice (which have their PML gene deleted) cannot assemble nuclear bodies, develop normally and live well, demonstrating that PML bodies are dispensable for most basic biological functions.

Splicing Speckles

Speckles are subnuclear structures that are enriched in pre-messenger RNA splicing factors and are located in the interchromatin regions of the nucleoplasm of mammalian cells. At the fluorescence-microscope level they appear as irregular, punctate structures, which vary in size and shape, and when examined by electron microscopy they are seen as clusters of interchromatin granules. Speckles are dynamic structures, and both their protein and RNA-protein components can cycle continuously between speckles and other nuclear locations, including active transcription sites. Studies on the composition, structure and behaviour of speckles have provided a model for understanding the functional compartmentalization of the nucleus and the organization of the gene-expression machinery splicing snRNPs and other splicing proteins necessary for pre-mRNA processing. Because of a cell's changing requirements, the composition and location of these bodies changes according to mRNA transcription and regulation via phosphorylation of specific proteins. The splicing speckles are also known as nuclear speckles (nuclear specks), splicing factor compartments (SF compartments), interchromatin granule clusters (IGCs), B snurposomes. B snurposomes are found in the amphibian oocyte nuclei and in *Drosophila melanogaster* embryos. B snurposomes appear alone or attached to the Cajal bodies in the electron micrographs of the amphibian nuclei. IGCs function as storage sites for the splicing factors.

Paraspeckles

Discovered by Fox et al. in 2002, paraspeckles are irregularly shaped compartments in the nucleus' interchromatin space. First documented in HeLa cells, where there are generally 10–30 per nucleus, paraspeckles are now known to also exist in all human primary cells, transformed cell lines, and tissue sections. Their name is derived from their distribution in the nucleus; the "para" is short for parallel and the "speckles" refers to the splicing speckles to which they are always in close proximity.

Paraspeckles are dynamic structures that are altered in response to changes in cellular metabolic activity. They are transcription dependent and in the absence of RNA Pol II transcription, the paraspeckle disappears and all of its associated protein components (PSP1, p54nrb, PSP2, CFI(m)68, and PSF) form a crescent shaped perinucleolar cap in the nucleolus. This phenomenon is demonstrated during the cell cycle. In the cell cycle, paraspeckles are present during interphase and during all of mitosis except for telophase. During telophase, when the two daughter nuclei are formed, there is no RNA Pol II transcription so the protein components instead form a perinucleolar cap.

Perichromatin Fibrils

Perichromatin fibrils are visible only under electron microscope. They are located next to the transcriptionally active chromatin and are hypothesized to be the sites of active pre-mRNA processing.

Function

The nucleus provides a site for genetic transcription that is segregated from the location of translation in the cytoplasm, allowing levels of gene regulation that are not available to prokaryotes. The main function of the cell nucleus is to control gene expression and mediate the replication of DNA during the cell cycle.

The nucleus is an organelle found in eukaryotic cells. Inside its fully enclosed nuclear membrane, it contains the majority of the cell's genetic material. This material is organized as DNA molecules, along with a variety of proteins, to form chromosomes.

Cell Compartmentalization

The nuclear envelope allows the nucleus to control its contents, and separate them from the rest of the cytoplasm where necessary. This is important for controlling processes on either side of the nuclear membrane. In most cases where a cytoplasmic process needs to be restricted, a key participant is removed to the nucleus, where it interacts with transcription factors to downregulate the production of certain enzymes in the pathway. This regulatory mechanism occurs in the case of glycolysis, a cellular pathway for breaking down glucose to produce energy. Hexokinase is an enzyme responsible for the first the step of glycolysis, forming glucose-6-phosphate from glucose. At high concentrations of fructose-6-phosphate, a molecule made later from glucose-6-phosphate, a regulator protein removes hexokinase to the nucleus, where it forms a transcriptional repressor complex with nuclear proteins to reduce the expression of genes involved in glycolysis.

In order to control which genes are being transcribed, the cell separates some transcription factor proteins responsible for regulating gene expression from physical access to the DNA until they are activated by other signaling pathways. This prevents even low levels of inappropriate gene expression. For example, in the case of NF-κB-controlled genes, which are involved in most inflammatory responses, transcription is induced in response to a signal pathway such as that initiated by the signaling molecule TNF-α, binds to a cell membrane receptor, resulting in the recruitment of signalling proteins, and eventually activating the transcription factor NF-κB. A nuclear localisation signal on the NF-κB protein allows it to be transported through the nuclear pore and into the nucleus, where it stimulates the transcription of the target genes.

The compartmentalization allows the cell to prevent translation of unspliced mRNA. Eukaryotic mRNA contains introns that must be removed before being translated to produce functional proteins. The splicing is done inside the nucleus before the mRNA can be accessed by ribosomes for translation. Without the nucleus, ribosomes would translate newly transcribed (unprocessed) mRNA, resulting in malformed and nonfunctional proteins.

Gene Expression

Gene expression first involves transcription, in which DNA is used as a template to produce RNA. In the case of genes encoding proteins, that RNA produced from this process is messenger RNA (mRNA), which then needs to be translated by ribosomes to form a protein. As ribosomes are located outside the nucleus, mRNA produced needs to be exported.

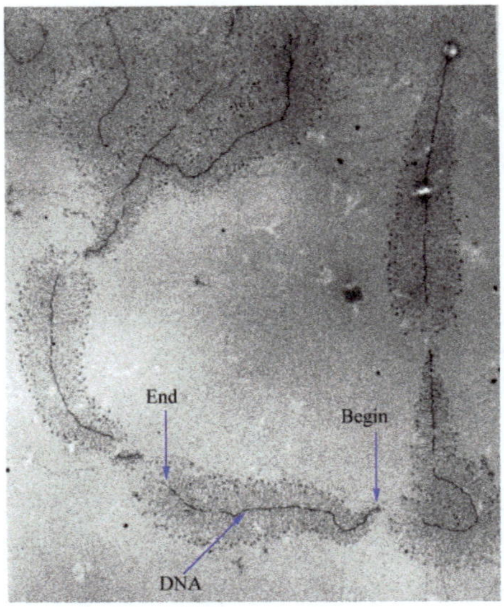

A micrograph of ongoing gene transcription of ribosomal RNA illustrating the growing primary transcripts. "Begin" indicates the 5' end of the DNA, where new RNA synthesis begins; "end" indicates the 3' end, where the primary transcripts are almost complete.

Since the nucleus is the site of transcription, it also contains a variety of proteins that either directly mediate transcription or are involved in regulating the process. These proteins include helicases, which unwind the double-stranded DNA molecule to facilitate access to it, RNA polymerases, which synthesize the growing RNA molecule, topoisomerases, which change the amount of supercoiling in DNA, helping it wind and unwind, as well as a large variety of transcription factors that regulate expression.

Processing of pre-MRNA

Newly synthesized mRNA molecules are known as primary transcripts or pre-mRNA. They must undergo post-transcriptional modification in the nucleus before being exported to the cytoplasm; mRNA that appears in the cytoplasm without these modifications is degraded rather than used for protein translation. The three main modifications are 5' capping, 3' polyadenylation, and RNA splicing. While in the nucleus, pre-mRNA is associated with a variety of proteins in complexes known as heterogeneous ribonucleoprotein particles (hnRNPs). Addition of the 5' cap occurs co-transcriptionally and is the first step in post-transcriptional modification. The 3' poly-adenine tail is only added after transcription is complete.

RNA splicing, carried out by a complex called the spliceosome, is the process by which introns, or regions of DNA that do not code for protein, are removed from the pre-mRNA and the remaining exons connected to re-form a single continuous molecule. This process normally occurs after 5' capping and 3' polyadenylation but can begin before synthesis is complete in transcripts with many exons. Many pre-mRNAs, including those encoding antibodies, can be spliced in multiple ways to produce different mature mRNAs that encode different protein sequences. This process is known as alternative splicing, and allows production of a large variety of proteins from a limited amount of DNA.

Dynamics and Regulation

Nuclear Transport

The entry and exit of large molecules from the nucleus is tightly controlled by the nuclear pore complexes. Although small molecules can enter the nucleus without regulation, macromolecules such as RNA and proteins require association karyopherins called importins to enter the nucleus and exportins to exit. "Cargo" proteins that must be translocated from the cytoplasm to the nucleus contain short amino acid sequences known as nuclear localization signals, which are bound by importins, while those transported from the nucleus to the cytoplasm carry nuclear export signals bound by exportins. The ability of importins and exportins to transport their cargo is regulated by GTPases, enzymes that hydrolyze the molecule guanosine triphosphate to release energy. The key GTPase in nuclear transport is Ran, which can bind either GTP or GDP (guanosine diphosphate), depending on whether it is located in the nucleus or the cytoplasm. Whereas importins depend on RanGTP to dissociate from their cargo, exportins require RanGTP in order to bind to their cargo

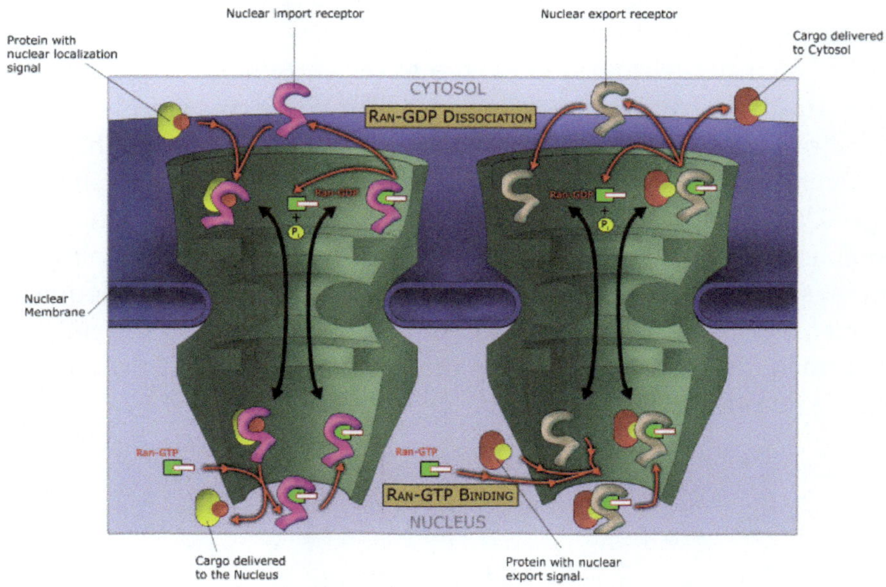

Macromolecules, such as RNA and proteins, are actively transported across the nuclear membrane in a process called the Ran-GTP nuclear transport cycle.

Nuclear import depends on the importin binding its cargo in the cytoplasm and carrying it through the nuclear pore into the nucleus. Inside the nucleus, RanGTP acts to separate the cargo from the importin, allowing the importin to exit the nucleus and be reused. Nuclear export is similar, as the exportin binds the cargo inside the nucleus in a process facilitated by RanGTP, exits through the nuclear pore, and separates from its cargo in the cytoplasm.

Specialized export proteins exist for translocation of mature mRNA and tRNA to the cytoplasm after post-transcriptional modification is complete. This quality-control mechanism is important due to these molecules' central role in protein translation. Mis-expression of a protein due to incomplete excision of exons or mis-incorporation of amino acids could have negative consequences for the cell; thus, incompletely modified RNA that reaches the cytoplasm is degraded rather than used in translation.

Assembly and Disassembly

During its lifetime, a nucleus may be broken down or destroyed, either in the process of cell division or as a consequence of apoptosis (the process of programmed cell death). During these events, the structural components of the nucleus — the envelope and lamina — can be systematically degraded. In most cells, the disassembly of the nuclear envelope marks the end of the prophase of mitosis. However, this disassembly of the nucleus is not a universal feature of mitosis and does not occur in all cells. Some unicellular eukaryotes (e.g., yeasts) undergo so-called closed mitosis, in which the nuclear envelope remains intact. In closed mitosis, the daughter chromosomes migrate to opposite poles of the nucleus, which then divides in two. The cells of higher eukaryotes, however, usually undergo open mitosis, which is characterized by breakdown of the nuclear envelope. The daughter chromosomes then migrate to opposite poles of the mitotic spindle, and new nuclei reassemble around them.

At a certain point during the cell cycle in open mitosis, the cell divides to form two cells. In order for this process to be possible, each of the new daughter cells must have a full set of genes, a process requiring replication of the chromosomes as well as segregation of the separate sets. This occurs by the replicated chromosomes, the sister chromatids, attaching to microtubules, which in turn are attached to different centrosomes. The sister chromatids can then be pulled to separate locations in the cell. In many cells, the centrosome is located in the cytoplasm, outside the nucleus; the microtubules would be unable to attach to the chromatids in the presence of the nuclear envelope. Therefore, the early stages in the cell cycle, beginning in prophase and until around prometaphase, the nuclear membrane is dismantled. Likewise, during the same period, the nuclear lamina is also disassembled, a process regulated by phosphorylation of the lamins by protein kinases such as the CDC2 protein kinase. Towards the end of the cell cycle, the nuclear membrane is reformed, and around the same time, the nuclear lamina are reassembled by dephosphorylating the lamins.

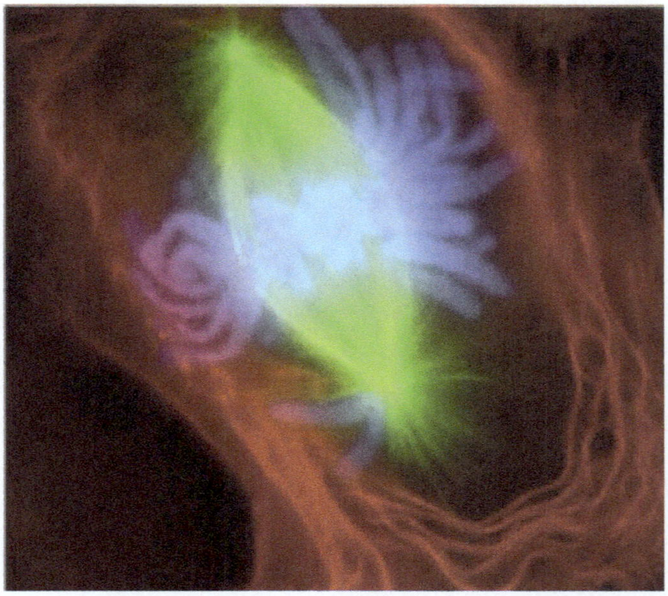

An image of a newt lung cell stained with fluorescent dyes during metaphase. The mitotic spindle can be seen, stained green, attached to the two sets of chromosomes, stained light blue. All chromosomes but one are already at the metaphase plate.

However, in dinoflagellates, the nuclear envelope remains intact, the centrosomes are located in the cytoplasm, and the microtubules come in contact with chromosomes, whose centromeric regions are incorporated into the nuclear envelope (the so-called closed mitosis with extranuclear spindle). In many other protists (e.g., ciliates, sporozoans) and fungi, the centrosomes are intranuclear, and their nuclear envelope also does not disassemle during cell division.

Apoptosis is a controlled process in which the cell's structural components are destroyed, resulting in death of the cell. Changes associated with apoptosis directly affect the nucleus and its contents, for example, in the condensation of chromatin and the disintegration of the nuclear envelope and lamina. The destruction of the lamin networks is controlled by specialized apoptotic proteases called caspases, which cleave the lamin proteins and, thus, degrade the nucleus' structural integrity. Lamin cleavage is sometimes used as a laboratory indicator of caspase activity in assays for early apoptotic activity. Cells that express mutant caspase-resistant lamins are deficient in nuclear changes related to apoptosis, suggesting that lamins play a role in initiating the events that lead to apoptotic degradation of the nucleus. Inhibition of lamin assembly itself is an inducer of apoptosis.

The nuclear envelope acts as a barrier that prevents both DNA and RNA viruses from entering the nucleus. Some viruses require access to proteins inside the nucleus in order to replicate and/or assemble. DNA viruses, such as herpesvirus replicate and assemble in the cell nucleus, and exit by budding through the inner nuclear membrane. This process is accompanied by disassembly of the lamina on the nuclear face of the inner membrane.

Disease-Related Dynamics

Initially, it has been suspected that immunoglobulins in general and autoantibodies in particular do not enter the nucleus. Now there is a body of evidence that under pathological conditions (e.g. lupus erythematosus) IgG can enter the nucleus.

Nuclei Per Cell

Most eukaryotic cell types usually have a single nucleus, but some have no nuclei, while others have several. This can result from normal development, as in the maturation of mammalian red blood cells, or from faulty cell division.

Anucleated Cells

An anucleated cell contains no nucleus and is, therefore, incapable of dividing to produce daughter cells. The best-known anucleated cell is the mammalian red blood cell, or erythrocyte, which also lacks other organelles such as mitochondria, and serves primarily as a transport vessel to ferry oxygen from the lungs to the body's tissues. Erythrocytes mature through erythropoiesis in the bone marrow, where they lose their nuclei, organelles, and ribosomes. The nucleus is expelled during the process of differentiation from an erythroblast to a reticulocyte, which is the immediate precursor of the mature erythrocyte. The presence of mutagens may induce the release of some immature "micronucleated" erythrocytes into the bloodstream. Anucleated cells can also arise from flawed cell division in which one daughter lacks a nucleus and the other has two nuclei.

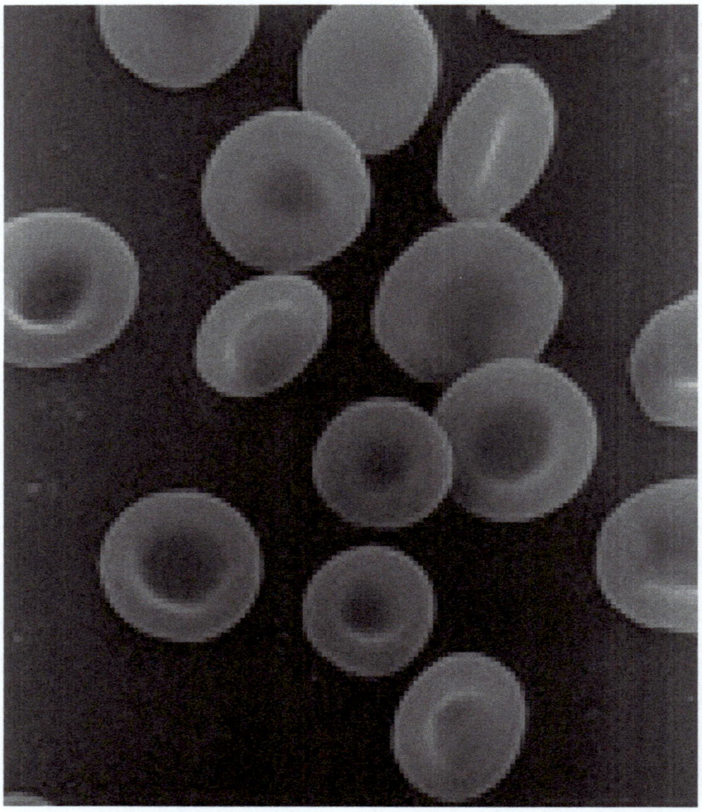

Human red blood cells, like those of other mammals, lack nuclei. This occurs as a normal part of the cells' development.

In flowering plants, this condition occurs in sieve tube elements.

Multinucleated Cells

Multinucleated cells contain multiple nuclei. Most acantharean species of protozoa and some fungi in mycorrhizae have naturally multinucleated cells. Other examples include the intestinal parasites in the genus *Giardia*, which have two nuclei per cell. In humans, skeletal muscle cells, called myocytes and syncytium, become multinucleated during development; the resulting arrangement of nuclei near the periphery of the cells allows maximal intracellular space for myofibrils. Multinucleated and binucleated cells can also be abnormal in humans; for example, cells arising from the fusion of monocytes and macrophages, known as giant multinucleated cells, sometimes accompany inflammation and are also implicated in tumor formation.

A number of dinoflagelates are known to have two nuclei. Unlike other multinucleated cells these nuclei contain two distinct lineages of DNA: one from the dinoflagelate and the other from a symbiotic diatom. Curiously the mitochondrion and the plastid of the diatom remain functional.

Evolution

As the major defining characteristic of the eukaryotic cell, the nucleus' evolutionary origin has been the subject of much speculation. Four major hypotheses have been proposed to explain the existence of the nucleus, although none have yet earned widespread support.

The first model known as the "syntrophic model" proposes that a symbiotic relationship between the archaea and bacteria created the nucleus-containing eukaryotic cell. (Organisms of the Archaea and Bacteria domain have no cell nucleus.) It is hypothesized that the symbiosis originated when ancient archaea, similar to modern methanogenic archaea, invaded and lived within bacteria similar to modern myxobacteria, eventually forming the early nucleus. This theory is analogous to the accepted theory for the origin of eukaryotic mitochondria and chloroplasts, which are thought to have developed from a similar endosymbiotic relationship between proto-eukaryotes and aerobic bacteria. The archaeal origin of the nucleus is supported by observations that archaea and eukarya have similar genes for certain proteins, including histones. Observations that myxobacteria are motile, can form multicellular complexes, and possess kinases and G proteins similar to eukarya, support a bacterial origin for the eukaryotic cell.

A second model proposes that proto-eukaryotic cells evolved from bacteria without an endosymbiotic stage. This model is based on the existence of modern planctomycetes bacteria that possess a nuclear structure with primitive pores and other compartmentalized membrane structures. A similar proposal states that a eukaryote-like cell, the chronocyte, evolved first and phagocytosed archaea and bacteria to generate the nucleus and the eukaryotic cell.

The most controversial model, known as *viral eukaryogenesis*, posits that the membrane-bound nucleus, along with other eukaryotic features, originated from the infection of a prokaryote by a virus. The suggestion is based on similarities between eukaryotes and viruses such as linear DNA strands, mRNA capping, and tight binding to proteins (analogizing histones to viral envelopes). One version of the proposal suggests that the nucleus evolved in concert with phagocytosis to form an early cellular "predator". Another variant proposes that eukaryotes originated from early archaea infected by poxviruses, on the basis of observed similarity between the DNA polymerases in modern poxviruses and eukaryotes. It has been suggested that the unresolved question of the evolution of sex could be related to the viral eukaryogenesis hypothesis.

A more recent proposal, the *exomembrane hypothesis*, suggests that the nucleus instead originated from a single ancestral cell that evolved a second exterior cell membrane; the interior membrane enclosing the original cell then became the nuclear membrane and evolved increasingly elaborate pore structures for passage of internally synthesized cellular components such as ribosomal subunits.

Cell Membrane

The cell membrane (also known as the plasma membrane or cytoplasmic membrane) is a biological membrane that separates the interior of all cells from the outside environment. The cell membrane is selectively permeable to ions and organic molecules and controls the movement of substances in and out of cells. The basic function of the cell membrane is to protect the cell from its surroundings.

It consists of the phospholipid bilayer with embedded proteins. Cell membranes are involved in a variety of cellular processes such as cell adhesion, ion conductivity and cell signalling and serve as the attachment surface for several extracellular structures, including the cell wall, glycocalyx, and intracellular cytoskeleton. Cell membranes can be artificially reassembled.

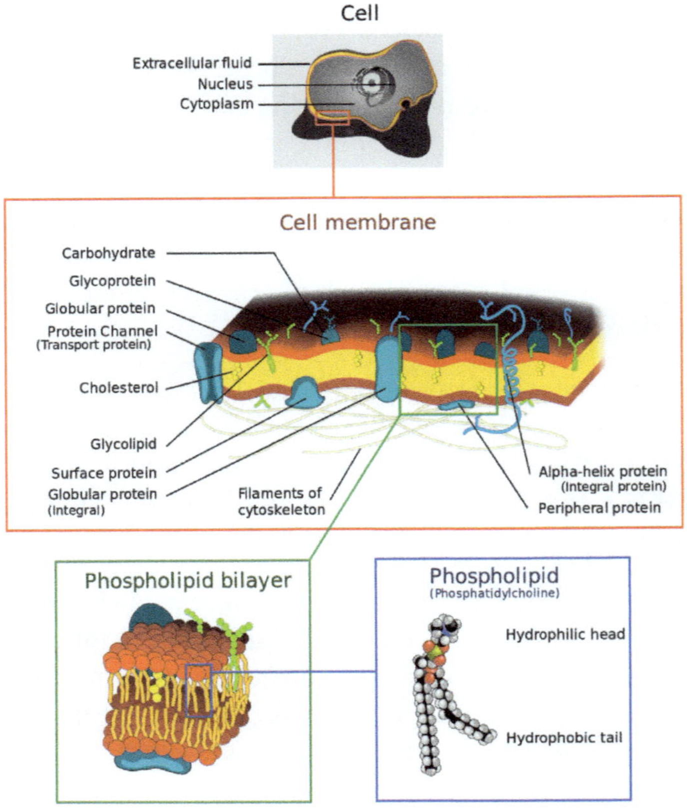

Illustration of a Eukaryotic cell membrane

Function

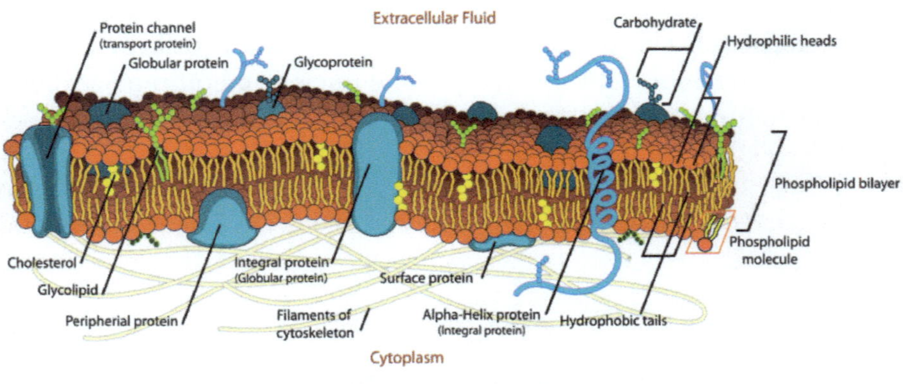

A detailed diagram of the cell membrane

The cell membrane (or plasma membrane or plasmalemma) surrounds the cytoplasm of living cells, physically separating the intracellular components from the extracellular environment. Fungi, bacteria and plants have a cell wall in addition, which provides a mechanical support to the cell and precludes the passage of larger molecules. The cell membrane also plays a role in anchoring the cytoskeleton to provide shape to the cell, and in attaching to the extracellular matrix and other cells to hold them together to form tissues.

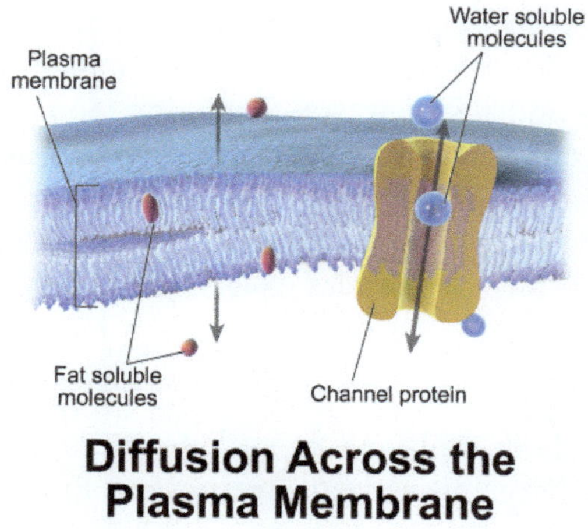

Illustration depicting cellular diffusion

The cell membrane is selectively permeable and able to regulate what enters and exits the cell, thus facilitating the transport of materials needed for survival. The movement of substances across the membrane can be either "passive", occurring without the input of cellular energy, or "active", requiring the cell to expend energy in transporting it. The membrane also maintains the cell potential. The cell membrane thus works as a selective filter that allows only certain things to come inside or go outside the cell. The cell employs a number of transport mechanisms that involve biological membranes:

1. Passive osmosis and diffusion: Some substances (small molecules, ions) such as carbon dioxide (CO_2) and oxygen (O_2), can move across the plasma membrane by diffusion, which is a passive transport process. Because the membrane acts as a barrier for certain molecules and ions, they can occur in different concentrations on the two sides of the membrane. Such a concentration gradient across a semipermeable membrane sets up an osmotic flow for the water.

2. Transmembrane protein channels and transporters: Nutrients, such as sugars or amino acids, must enter the cell, and certain products of metabolism must leave the cell. Such molecules diffuse passively through protein channels such as aquaporins (in the case of water (H_2O)) in facilitated diffusion or are pumped across the membrane by transmembrane transporters. Protein channel proteins, also called *permeases*, are usually quite specific, recognizing and transporting only a limited food group of chemical substances, often even only a single substance.

3. Endocytosis: Endocytosis is the process in which cells absorb molecules by engulfing them. The plasma membrane creates a small deformation inward, called an invagination, in which the substance to be transported is captured. The deformation then pinches off from the membrane on the inside of the cell, creating a vesicle containing the captured substance. Endocytosis is a pathway for internalizing solid particles ("cell eating" or phagocytosis), small molecules and ions ("cell drinking" or pinocytosis), and macromolecules. Endocytosis requires energy and is thus a form of active transport.

4. Exocytosis: Just as material can be brought into the cell by invagination and formation of a vesicle, the membrane of a vesicle can be fused with the plasma membrane, extruding its contents

to the surrounding medium. This is the process of exocytosis. Exocytosis occurs in various cells to remove undigested residues of substances brought in by endocytosis, to secrete substances such as hormones and enzymes, and to transport a substance completely across a cellular barrier. In the process of exocytosis, the undigested waste-containing food vacuole or the secretory vesicle budded from Golgi apparatus, is first moved by cytoskeleton from the interior of the cell to the surface. The vesicle membrane comes in contact with the plasma membrane. The lipid molecules of the two bilayers rearrange themselves and the two membranes are, thus, fused. A passage is formed in the fused membrane and the vesicles discharges its contents outside the cell.

Prokaryotes

Gram-negative bacteria have both a plasma membrane and an outer membrane separated by periplasm. Other prokaryotes have only a plasma membrane. Prokaryotic cells are also surrounded by a cell wall composed of peptidoglycan (amino acids and sugars). Some eukaryotic cells also have cell walls, but none that are made of peptidoglycan. The outer membrane of gram negative microbes is rich in lipopolysaccharide and thus is different from cell membrane of the microbes. The outer membrane can bleb out into periplasmic protrusions under stess conditions or upon virulence requirements while encountering a host target cell, and thus such blebs may work as virulence organelles.

Structures

Fluid Mosaic Model

According to the fluid mosaic model of S. J. Singer and G. L. Nicolson (1972), which replaced the earlier model of Davson and Danielli, biological membranes can be considered as a two-dimensional liquid in which lipid and protein molecules diffuse more or less easily. Although the lipid bilayers that form the basis of the membranes do indeed form two-dimensional liquids by themselves, the plasma membrane also contains a large quantity of proteins, which provide more structure. Examples of such structures are protein-protein complexes, pickets and fences formed by the actin-based cytoskeleton, and potentially lipid rafts.

Lipid Bilayer

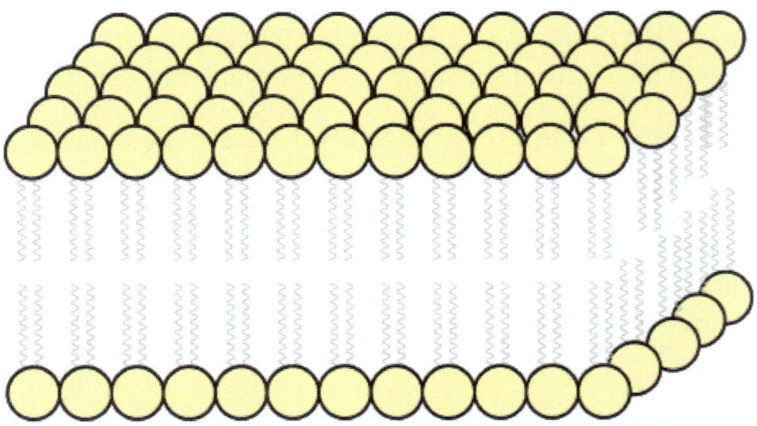

Diagram of the arrangement of amphipathic lipid molecules to form a lipid bilayer. The yellow polar head groups separate the grey hydrophobic tails from the aqueous cytosolic and extracellular environments.

Lipid bilayers form through the process of self-assembly. The cell membrane consists primarily of a thin layer of amphipathic phospholipids that spontaneously arrange so that the hydrophobic "tail" regions are isolated from the surrounding water while the hydrophilic "head" regions interact with the intracellular (cytosolic) and extracellular faces of the resulting bilayer. This forms a continuous, spherical lipid bilayer. Hydrophobic interactions (also known as the hydrophobic effect) are the major driving forces in the formation of lipid bilayers. An increase in interactions between hydrophobic molecules (causing clustering of hydrophobic regions) allows water molecules to bond more freely with each other, increasing the entropy of the system. This complex interaction can include noncovalent interactions such as van der Waals, electrostatic and hydrogen bonds.

Lipid bilayers are generally impermeable to ions and polar molecules. The arrangement of hydrophilic heads and hydrophobic tails of the lipid bilayer prevent polar solutes (ex. amino acids, nucleic acids, carbohydrates, proteins, and ions) from diffusing across the membrane, but generally allows for the passive diffusion of hydrophobic molecules. This affords the cell the ability to control the movement of these substances via transmembrane protein complexes such as pores, channels and gates.

Flippases and scramblases concentrate phosphatidyl serine, which carries a negative charge, on the inner membrane. Along with NANA, this creates an extra barrier to charged moieties moving through the membrane.

Membranes serve diverse functions in eukaryotic and prokaryotic cells. One important role is to regulate the movement of materials into and out of cells. The phospholipid bilayer structure (fluid mosaic model) with specific membrane proteins accounts for the selective permeability of the membrane and passive and active transport mechanisms. In addition, membranes in prokaryotes and in the mitochondria and chloroplasts of eukaryotes facilitate the synthesis of ATP through chemiosmosis.

Membrane Polarity

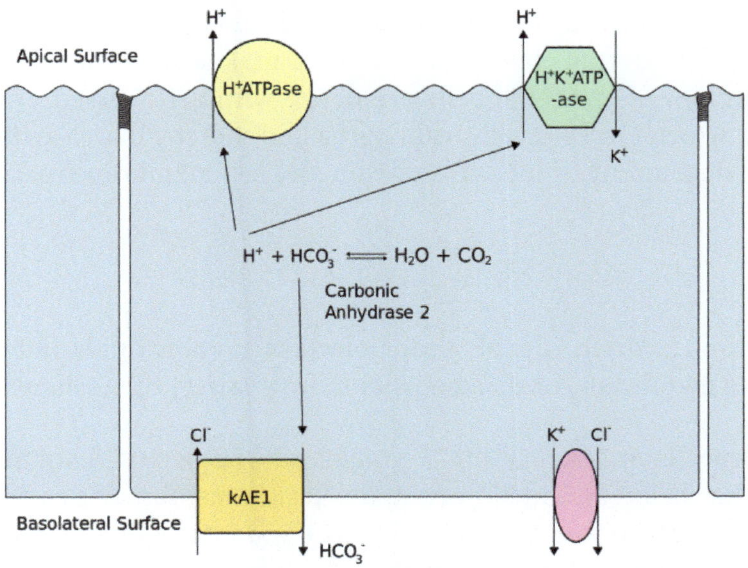

Alpha intercalated cell

The apical membrane of a polarized cell is the surface of the plasma membrane that faces inward to the lumen. This is particularly evident in epithelial and endothelial cells, but also describes other polarized cells, such as neurons. The basolateral membrane of a polarized cell is the surface of the plasma membrane that forms its basal and lateral surfaces. It faces outwards, towards the interstitium, and away from the lumen. Basolateral membrane is a compound phrase referring to the terms "basal (base) membrane" and "lateral (side) membrane", which, especially in epithelial cells, are identical in composition and activity. Proteins (such as ion channels and pumps) are free to move from the basal to the lateral surface of the cell or vice versa in accordance with the fluid mosaic model. Tight junctions join epithelial cells near their apical surface to prevent the migration of proteins from the basolateral membrane to the apical membrane. The basal and lateral surfaces thus remain roughly equivalent to one another, yet distinct from the apical surface.

Membrane Structures

Cell membrane can form different types of "supramembrane" structures such as caveola, postsynaptic density, podosome, invadopodium, focal adhesion, and different types of cell junctions. These structures are usually responsible for cell adhesion, communication, endocytosis and exocytosis. They can be visualized by electron microscopy or fluorescence microscopy. They are composed of specific proteins, such as integrins and cadherins.

Cytoskeleton

The cytoskeleton is found underlying the cell membrane in the cytoplasm and provides a scaffolding for membrane proteins to anchor to, as well as forming organelles that extend from the cell. Indeed, cytoskeletal elements interact extensively and intimately with the cell membrane. Anchoring proteins restricts them to a particular cell surface — for example, the apical surface of epithelial cells that line the vertebrate gut — and limits how far they may diffuse within the bilayer. The cytoskeleton is able to form appendage-like organelles, such as cilia, which are microtubule-based extensions covered by the cell membrane, and filopodia, which are actin-based extensions. These extensions are ensheathed in membrane and project from the surface of the cell in order to sense the external environment and/or make contact with the substrate or other cells. The apical surfaces of epithelial cells are dense with actin-based finger-like projections known as microvilli, which increase cell surface area and thereby increase the absorption rate of nutrients. Localized decoupling of the cytoskeleton and cell membrane results in formation of a bleb.

Composition

Cell membranes contain a variety of biological molecules, notably lipids and proteins. Material is incorporated into the membrane, or deleted from it, by a variety of mechanisms:

- Fusion of intracellular vesicles with the membrane (exocytosis) not only excretes the contents of the vesicle but also incorporates the vesicle membrane's components into the cell membrane. The membrane may form blebs around extracellular material that pinch off to become vesicles (endocytosis).

- If a membrane is continuous with a tubular structure made of membrane material, then material from the tube can be drawn into the membrane continuously.

- Although the concentration of membrane components in the aqueous phase is low (stable membrane components have low solubility in water), there is an exchange of molecules between the lipid and aqueous phases.

Lipids

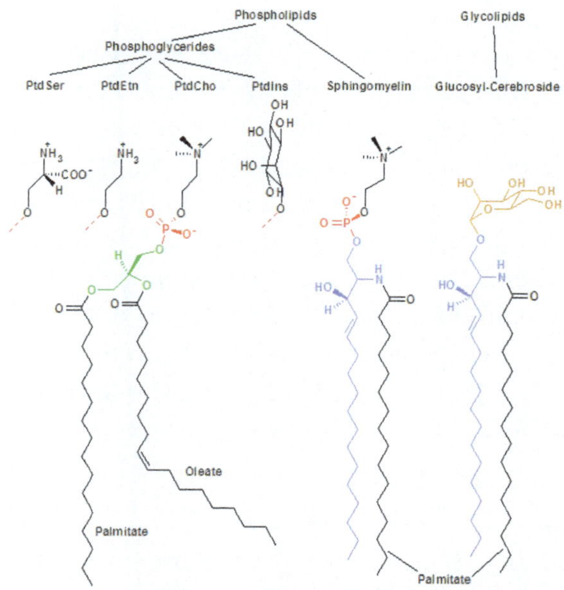

Examples of the major membrane phospholipids and glycolipids: phosphatidylcholine (PtdCho), phosphatidylethanolamine (PtdEtn), phosphatidylinositol (PtdIns), phosphatidylserine (PtdSer).

The cell membrane consists of three classes of amphipathic lipids: phospholipids, glycolipids, and sterols. The amount of each depends upon the type of cell, but in the majority of cases phospholipids are the most abundant. In RBC studies, 30% of the plasma membrane is lipid.

The fatty chains in phospholipids and glycolipids usually contain an even number of carbon atoms, typically between 16 and 20. The 16- and 18-carbon fatty acids are the most common. Fatty acids may be saturated or unsaturated, with the configuration of the double bonds nearly always "cis". The length and the degree of unsaturation of fatty acid chains have a profound effect on membrane fluidity as unsaturated lipids create a kink, preventing the fatty acids from packing together as tightly, thus decreasing the melting temperature (increasing the fluidity) of the membrane. The ability of some organisms to regulate the fluidity of their cell membranes by altering lipid composition is called homeoviscous adaptation.

The entire membrane is held together via non-covalent interaction of hydrophobic tails, however the structure is quite fluid and not fixed rigidly in place. Under physiological conditions phospholipid molecules in the cell membrane are in the liquid crystalline state. It means the lipid molecules are free to diffuse and exhibit rapid lateral diffusion along the layer in which they are present. However, the exchange of phospholipid molecules between intracellular and extracellular leaflets of the bilayer is a very slow process. Lipid rafts and caveolae are examples of cholesterol-enriched microdomains in the cell membrane. Also, a fraction of the lipid in direct contact with integral membrane proteins, which is tightly bound to the protein surface is called annular lipid shell; it behaves as a part of protein complex.

In animal cells cholesterol is normally found dispersed in varying degrees throughout cell membranes, in the irregular spaces between the hydrophobic tails of the membrane lipids, where it confers a stiffening and strengthening effect on the membrane.

Phospholipids Forming Lipid Vesicles

Lipid vesicles or liposomes are circular pockets that are enclosed by a lipid bilayer. These structures are used in laboratories to study the effects of chemicals in cells by delivering these chemicals directly to the cell, as well as getting more insight into cell membrane permeability. Lipid vesicles and liposomes are formed by first suspending a lipid in an aqueous solution then agitating the mixture through sonication, resulting in a vesicle. By measuring the rate of efflux from that of the inside of the vesicle to the ambient solution, allows researcher to better understand membrane permeability. Vesicles can be formed with molecules and ions inside the vesicle by forming the vesicle with the desired molecule or ion present in the solution. Proteins can also be embedded into the membrane through solubilizing the desired proteins in the presence of detergents and attaching them to the phospholipids in which the liposome is formed. These provide researchers with a tool to examine various membrane protein functions.

Carbohydrates

Plasma membranes also contain carbohydrates, predominantly glycoproteins, but with some glycolipids (cerebrosides and gangliosides). For the most part, no glycosylation occurs on membranes within the cell; rather generally glycosylation occurs on the extracellular surface of the plasma membrane. The glycocalyx is an important feature in all cells, especially epithelia with microvilli. Recent data suggest the glycocalyx participates in cell adhesion, lymphocyte homing, and many others. The penultimate sugar is galactose and the terminal sugar is sialic acid, as the sugar backbone is modified in the Golgi apparatus. Sialic acid carries a negative charge, providing an external barrier to charged particles.

Proteins

Type	Description	Examples
Integral proteins or transmembrane proteins	Span the membrane and have a hydrophilic cytosolic domain, which interacts with internal molecules, a hydrophobic membrane-spanning domain that anchors it within the cell membrane, and a hydrophilic extracellular domain that interacts with external molecules. The hydrophobic domain consists of one, multiple, or a combination of α-helices and β sheet protein motifs.	Ion channels, proton pumps, G protein-coupled receptor

Lipid anchored proteins	Covalently bound to single or multiple lipid molecules; hydrophobically insert into the cell membrane and anchor the protein. The protein itself is not in contact with the membrane.	G proteins
Peripheral proteins	Attached to integral membrane proteins, or associated with peripheral regions of the lipid bilayer. These proteins tend to have only temporary interactions with biological membranes, and once reacted, the molecule dissociates to carry on its work in the cytoplasm.	Some enzymes, some hormones

The cell membrane has large content of proteins, typically around 50% of membrane volume These proteins are important for cell because they are responsible for various biological activities. Approximately a third of the genes in yeast code specifically for them, and this number is even higher in multicellular organisms.

The cell membrane, being exposed to the outside environment, is an important site of cell–cell communication. As such, a large variety of protein receptors and identification proteins, such as antigens, are present on the surface of the membrane. Functions of membrane proteins can also include cell–cell contact, surface recognition, cytoskeleton contact, signaling, enzymatic activity, or transporting substances across the membrane.

Most membrane proteins must be inserted in some way into the membrane. For this to occur, an N-terminus "signal sequence" of amino acids directs proteins to the endoplasmic reticulum, which inserts the proteins into a lipid bilayer. Once inserted, the proteins are then transported to their final destination in vesicles, where the vesicle fuses with the target membrane.

Variation

The cell membrane has different lipid and protein compositions in distinct types of cells and may have therefore specific names for certain cell types:

- Sarcolemma in myocytes
- Oolemma in oocytes
- Axolemma in neuronal processes - axons
- Historically, the plasma membrane was also referred to as the plasmalemma

Permeability

The permeability of a membrane is the rate of passive diffusion of molecules through the membrane. These molecules are known as permeant molecules. Permeability depends mainly on the electric charge and polarity of the molecule and to a lesser extent the molar mass of the molecule. Due to the cell membrane's hydrophobic nature, small electrically neutral molecules pass through the membrane more easily than charged, large ones. The inability of charged molecules to pass through the cell membrane results in pH partition of substances throughout the fluid compartments of the body.

Cell Cortex

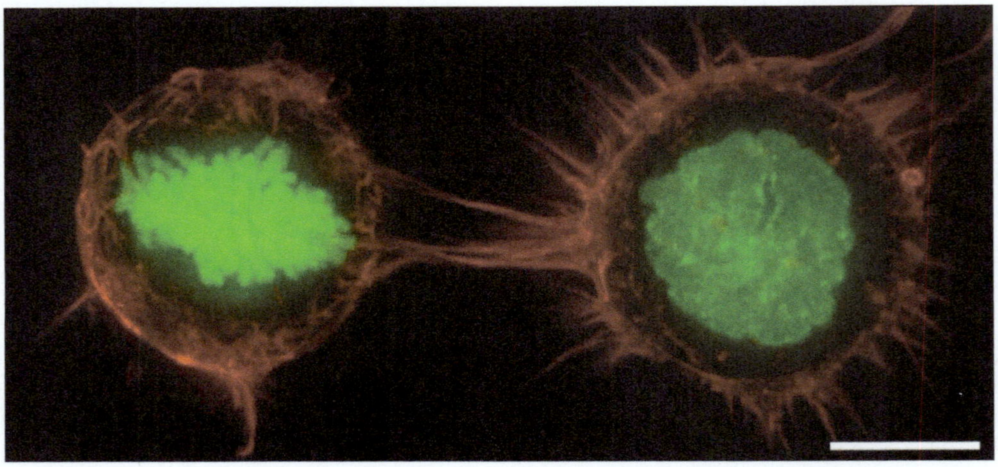

F-actin distribution in the cell cortex as shown by rhodamine phalloidin staining of HeLa cells that constitutively express Histone H2B-GFP to mark chromosomes. F-actin is thus red, while Histone H2B is displayed in green. The left hand cell is in mitosis, as demonstrated by chromosome condensation, while the right hand cell is in interphase(as determined by intact cell nucleus) in a suspended state. In both cases F-actin is enriched around the cell periphery. Scale bar: 10 micrometers.

The cell cortex, also known as the actin cortex or actomyosin cortex, is a specialized layer of cytoplasmic protein on the inner face of the plasma membrane of the cell periphery. It functions as a modulator of plasma membrane behavior and cell surface properties. In most eukaryotic cells lacking a cell wall, the cortex is an actin-rich network consisting of F-actin filaments, myosin motors, and actin-binding proteins. The actomyosin cortex is attached to the cell membrane via membrane-anchoring proteins called ERM proteins and it plays a central role in cell shape control. The protein constituents of the cortex undergo rapid turnover, making the cortex both mechanically rigid and highly plastic, two properties essential to its function. In most cases, the cortex is in the range of 100 to 1000 nanometers thick.

In some animal cells, the protein spectrin may be present in the cortex. Spectrin helps to create a network by cross-linked actin filaments. The proportions of spectrin and actin vary with cell type. Spectrin proteins and actin microfilaments are attached to transmembrane proteins by attachment proteins between them and the transmembrane proteins. The cell cortex is attached to the inner (cytosolic) face of the plasma membrane in cells where the spectrin proteins and actin microfilaments form a mesh-like structure much like a fishnet except that it can be broken and reformed.

In plant cells, the cell cortex is reinforced by cortical microtubules underlying the plasma membrane. The direction of these cortical microtubules determines which way the cell elongates when it grows.

Functions

- In mitosis, F-actin and myosin II form a highly contractile and uniform cortex to drive mitotic cell rounding. The surface tension produced by the actomyosin cortex activity generates intracellular hydrostatic pressure capable of displacing surrounding objects to facilitate rounding.
- In cytokinesis the cell cortex plays a central role by producing a myosin-rich contractile ring to constrict the dividing cell into two daughter cells.
- Cell cortex contractility is key for amoeboidal type cell migration characteristic of many cancer cell metastasis events.

Cell wall

A cell wall is a structural layer that surrounds some types of cells, situated outside the cell membrane. It can be tough, flexible, and sometimes rigid. It provides cells with both structural support and protection, and also acts as a filtering mechanism. Cell walls are present in plants, fungi and prokaryotic cells, where a major function is to act as pressure vessels, preventing over-expansion when water enters the cells. Cell walls are absent from mycoplasmas.

The composition of cell walls varies between species and may depend on cell type and developmental stage. The primary cell wall of land plants is composed of the polysaccharides cellulose, hemicellulose and pectin. In bacteria, the cell wall is composed of peptidoglycan. Archaean cell walls have various compositions, and may be formed of glycoprotein S-layers, pseudopeptidoglycan, or polysaccharides. Fungi possess cell walls made of the glucosamine polymer chitin, and algae typically possess walls made of glycoproteins and polysaccharides. Unusually, diatoms have a cell wall composed of biogenic silica. Often, other accessory molecules such as lignin or cutin are found anchored to the cell wall.

History

A plant cell wall was first observed and named (simply as a "wall") by Robert Hooke in 1665. However, "the dead excrusion product of the living protoplast" was forgotten, for almost three centuries, being subject of scientific interest mainly as a resource for industrial processing or in relation to animal or human health.

In 1804, Karl Rudolphi and J.H.F. Link proved that cells had independent cell walls. Before, it had been thought that cells shared walls and the fluid passed between them this way.

The mode of formation of the cell wal was controversial in the 19th century. Hugo von Mohl (1853, 1858) advocated the idea that the cell wall grows by apposition. Carl Nägeli (1858, 1862, 1863) believed that the growth of the wall in thickness and in area he was due to a process termed intussusception. The apposition, or lamination, theory was improved by Eduard Strasburger (1882,

1889), as the intussusception theory was by Julius Wiesner (1886).

In 1930, Münch would coin the term apoplast in order to separate the "living" symplast from the "dead" plant region.

By the 1980s, some authors suggested to replace the term "cell wall" of plants with the more precise term "extracellular matrix", used for animal cells, but others prefer to maintain the older term.

Properties

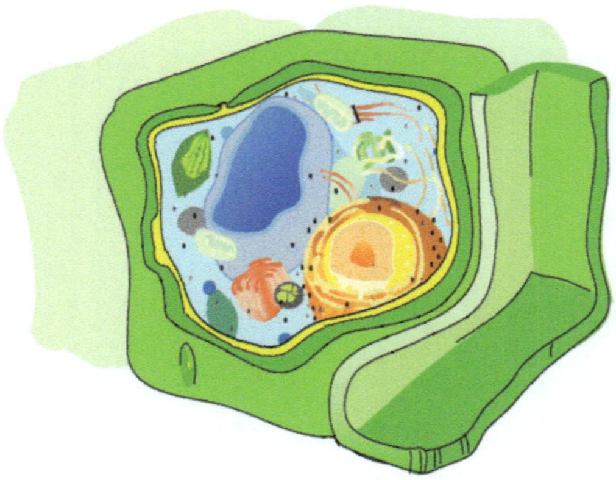

Diagram of the plant cell, with the cell wall in green.

Cell walls serve similar purposes in those organisms that possess them. They may give cells rigidity and strength, offering protection against mechanical stress. In multicellular organisms, they permit the organism to build and hold its shape (morphogenesis). Cell walls also limit the entry of large molecules that may be toxic to the cell. They further permit the creation of stable osmotic environments by preventing osmotic lysis and helping to retain water. Their composition, properties, and form may change during the cell cycle and depend on growth conditions.

Rigidity of Cell Walls

In most cells, the cell wall is flexible, meaning that it will bend rather than holding a fixed shape, but has considerable tensile strength. The apparent rigidity of primary plant tissues is enabled by cell walls, but is not due to the walls' stiffness. Hydraulic turgor pressure creates this rigidity, along with the wall structure. The flexibility of the cell walls is seen when plants wilt, so that the stems and leaves begin to droop, or in seaweeds that bend in water currents. As John Howland states it:

Think of the cell wall as a wicker basket in which a balloon has been inflated so that it exerts pressure from the inside. Such a basket is very rigid and resistant to mechanical damage. Thus does the prokaryote cell (and eukaryotic cell that possesses a cell wall) gain strength from a flexible plasma membrane pressing against a rigid cell wall.

The apparent rigidity of the cell wall thus results from inflation of the cell contained within. This inflation is a result of the passive uptake of water.

In plants, a secondary cell wall is a thicker additional layer of cellulose which increases wall rigidity. Additional layers may be formed containing lignin in xylem cell walls, or containing suberin in cork cell walls. These compounds are rigid and waterproof, making the secondary wall stiff. Both wood and bark cells of trees have secondary walls. Other parts of plants such as the leaf stalk may acquire similar reinforcement to resist the strain of physical forces.

Permeability

The primary cell wall of most plant cells is freely permeable to small molecules including small proteins, with size exclusion estimated to be 30-60 kDa.Key nutrients, especially water and carbon dioxide, are distributed throughout the plant from cell wall to cell wall in apoplastic flow. The pH is an important factor governing the transport of molecules through cell walls.

Cell Wall Evolution

Cell walls evolved independently in many groups, even in the photosynthetic eukaryotes. In these lineages, the cell wall is involved in the evolution of the multicellularity, terrestrialization and vascularization.

Plant Cell Walls

The walls of plant cells must have sufficient tensile strength to withstand internal osmotic pressures of several times atmospheric pressure that result from the difference in solute concentration between the cell interior and external water. Plant cell walls vary from 0.1 to several μm in thickness.

Layers

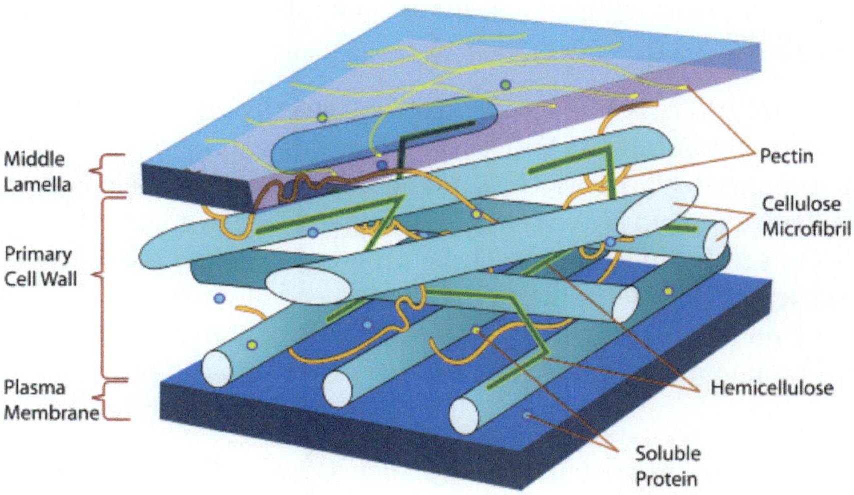

Molecular structure of the primary cell wall in plants.

Up to three strata or layers may be found in plant cell walls:

- The primary cell wall, generally a thin, flexible and extensible layer formed while the cell is growing.

- The secondary cell wall, a thick layer formed inside the primary cell wall after the cell is fully grown. It is not found in all cell types. Some cells, such as the conducting cells in xylem, possess a secondary wall containing lignin, which strengthens and waterproofs the wall.

- The middle lamella, a layer rich in pectins. This outermost layer forms the interface between adjacent plant cells and glues them together.

Composition

In the primary (growing) plant cell wall, the major carbohydrates are cellulose, hemicellulose and pectin. The cellulose microfibrils are linked via hemicellulosic tethers to form the cellulose-hemicellulose network, which is embedded in the pectin matrix. The most common hemicellulose in the primary cell wall is xyloglucan. In grass cell walls, xyloglucan and pectin are reduced in abundance and partially replaced by glucuronarabinoxylan, another type of hemicellulose. Primary cell walls characteristically extend (grow) by a mechanism called acid growth, which involves turgor-driven movement of the strong cellulose microfibrils within the weaker hemicellulose/pectin matrix, catalyzed by expansin proteins. The outer part of the primary cell wall of the plant epidermis is usually impregnated with cutin and wax, forming a permeability barrier known as the plant cuticle.

Secondary cell walls contain a wide range of additional compounds that modify their mechanical properties and permeability. The major polymers that make up wood (largely secondary cell walls) include:

- cellulose, 35-50%
- xylan, 20-35%, a type of hemicellulose
- lignin, 10-25%, a complex phenolic polymer that penetrates the spaces in the cell wall between cellulose, hemicellulose and pectin components, driving out water and strengthening the wall.

Additionally, structural proteins (1-5%) are found in most plant cell walls; they are classified as hydroxyproline-rich glycoproteins (HRGP), arabinogalactan proteins (AGP), glycine-rich proteins (GRPs), and proline-rich proteins (PRPs). Each class of glycoprotein is defined by a characteristic, highly repetitive protein sequence. Most are glycosylated, contain hydroxyproline (Hyp) and become cross-linked in the cell wall. These proteins are often concentrated in specialized cells and in cell corners. Cell walls of the epidermis and endodermis may also contain suberin or cutin, two polyester polymers that function as permeability barriers. The relative composition of carbohydrates, secondary compounds and protein varies between plants and between the cell type and age.

Plant cells walls also contain numerous enzymes, such as hydrolases, esterases, peroxidases, and transglycosylases, that cut, trim and cross-link wall polymers.

The walls of cork cells in the bark of trees are impregnated with suberin, and suberin also forms the permeability barrier in primary roots known as the Casparian strip. Secondary walls - especially in grasses - may also contain microscopic silica crystals, which may strengthen the wall and protect it from herbivores.

Cell walls in some plant tissues also function as storage depots for carbohydrates that can be broken down and resorbed to supply the metabolic and growth needs of the plant. For example, endosperm cell walls in the seeds of cereal grasses, nasturtium, and other species, are rich in glucans and other polysaccharides that are readily digested by enzymes during seed germination to form simple sugars that nourish the growing embryo. Cellulose microfibrils are not readily digested by plants, however.

Formation

The middle lamella is laid down first, formed from the cell plate during cytokinesis, and the primary cell wall is then deposited inside the middle lamella. The actual structure of the cell wall is not clearly defined and several models exist - the covalently linked cross model, the tether model, the diffuse layer model and the stratified layer model. However, the primary cell wall, can be defined as composed of cellulose microfibrils aligned at all angles. Cellulose microfibrils are produced at the plasma membrane by the cellulose synthase complex, which is proposed to be made of a hexameric rosette that contains three cellulose synthase catalytic subunits for each of the six units. Microfibrils are held together by hydrogen bonds to provide a high tensile strength. The cells are held together and share the gelatinous membrane called the *middle lamella*, which contains magnesium and calcium pectates (salts of pectic acid). Cells interact though plasmodesmata, which are inter-connecting channels of cytoplasm that connect to the protoplasts of adjacent cells across the cell wall.

In some plants and cell types, after a maximum size or point in development has been reached, a *secondary wall* is constructed between the plasma membrane and primary wall. Unlike the primary wall, the cellulose microfibrils are aligned parallel in layers, the orientation changing slightly with each additional layer so that the structure becomes helicoidal. Cells with secondary cell walls can be rigid, as in the gritty sclereid cells in pear and quince fruit. Cell to cell communication is possible through pits in the secondary cell wall that allow plasmodesmata to connect cells through the secondary cell walls.

Fungal Cell Walls

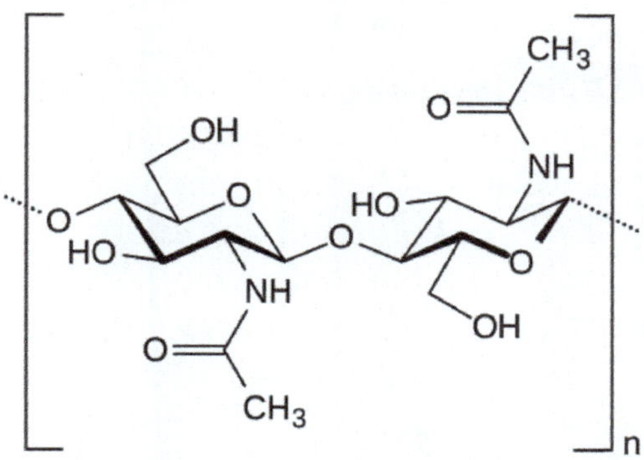

Chemical structure of a unit from a chitin polymer chain.

There are several groups of organisms that have been called “fungi”. Some of these groups (Oomycete and Myxogastria) have been transferred out of the Kingdom Fungi, in part because of fundamental biochemical differences in the composition of the cell wall. Most true fungi have a cell wall consisting largely of chitin and other polysaccharides. True fungi do not have cellulose in their cell walls.

True Fungi

In fungi, the cell wall is the outer-most layer, external to the plasma membrane. The fungal cell wall is a matrix of three main components:

- chitin: polymers consisting mainly of unbranched chains of β-(1,4)-linked-N-Acetylglucosamine in the Ascomycota and Basidiomycota, or poly-β-(1,4)-linked-N-Acetylglucosamine (chitosan) in the Zygomycota. Both chitin and chitosan are synthesized and extruded at the plasma membrane.
- glucans: glucose polymers that function to cross-link chitin or chitosan polymers. β-glucans are glucose molecules linked via β-(1,3)- or β-(1,6)- bonds and provide rigidity to the cell wall while α-glucans are defined by α-(1,3)- and/or α-(1,4) bonds and function as part of the matrix.
- proteins: enzymes necessary for cell wall synthesis and lysis in addition to structural proteins are all present in the cell wall. Most of the structural proteins found in the cell wall are glycosylated and contain mannose, thus these proteins are called mannoproteins or mannans.

Other Eukaryotic Cell Walls

Algae

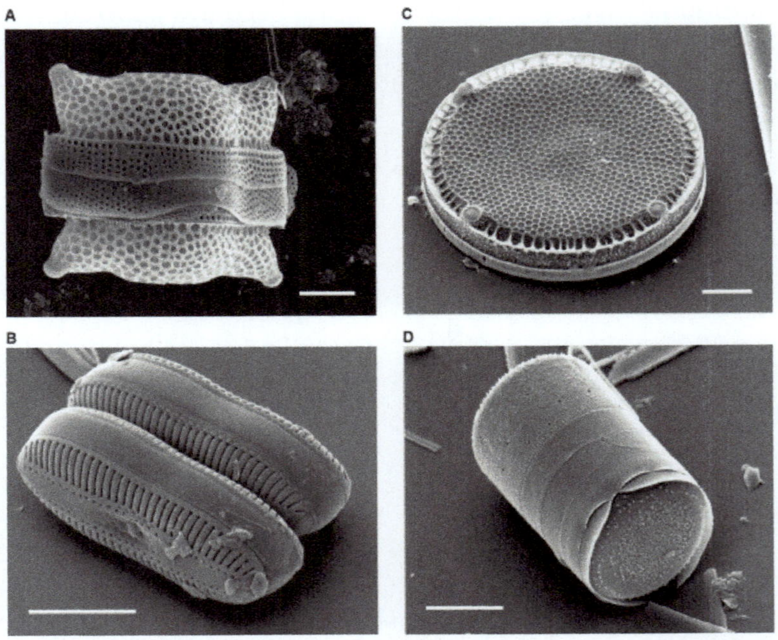

Scanning electron micrographs of diatoms showing the external appearance of the cell wall

Like plants, algae have cell walls. Algal cell walls contain either polysaccharides (such as cellulose (a glucan)) or a variety of glycoproteins (Volvocales) or both. The inclusion of additional polysaccharides in algal cells walls is used as a feature for algal taxonomy.

- Mannans: They form microfibrils in the cell walls of a number of marine green algae including those from the genera, *Codium*, *Dasycladus*, and *Acetabularia* as well as in the walls of some red algae, like *Porphyra* and *Bangia*.
- Xylans:
- Alginic acid: It is a common polysaccharide in the cell walls of brown algae.
- Sulfonated polysaccharides: They occur in the cell walls of most algae; those common in red algae include agarose, carrageenan, porphyran, furcelleran and funoran.

Other compounds that may accumulate in algal cell walls include sporopollenin and calcium ions.

The group of algae known as the diatoms synthesize their cell walls (also known as frustules or valves) from silicic acid (specifically orthosilicic acid, H_4SiO_4). The acid is polymerised intra-cellularly, then the wall is extruded to protect the cell. Significantly, relative to the organic cell walls produced by other groups, silica frustules require less energy to synthesize (approximately 8%), potentially a major saving on the overall cell energy budget and possibly an explanation for higher growth rates in diatoms.

In brown algae, phlorotannins may be a constituent of the cell walls.

Water Molds

The group Oomycetes, also known as water molds, are saprotrophic plant pathogens like fungi. Until recently they were widely believed to be fungi, but structural and molecular evidence has led to their reclassification as heterokonts, related to autotrophic brown algae and diatoms. Unlike fungi, oomycetes typically possess cell walls of cellulose and glucans rather than chitin, although some genera (such as *Achlya* and *Saprolegnia*) do have chitin in their walls. The fraction of cellulose in the walls is no more than 4 to 20%, far less than the fraction of glucans. Oomycete cell walls also contain the amino acid hydroxyproline, which is not found in fungal cell walls.

Slime Molds

The dictyostelids are another group formerly classified among the fungi. They are slime molds that feed as unicellular amoebae, but aggregate into a reproductive stalk and sporangium under certain conditions. Cells of the reproductive stalk, as well as the spores formed at the apex, possess a cellulose wall. The spore wall has been shown to possess three layers, the middle of which is composed primarily of cellulose, and the innermost is sensitive to cellulase and pronase.

Prokaryotic Cell Walls

Diagram of a typical Gram-positive bacterium. The cell envelope comprises a plasma membrane, seen here in green, and a thick peptidoglycan-containing cell wall (the yellow layer). No outer lipid membrane is present, as would be the case in Gram-negative bacteria. The red layer, known as the capsule, is distinct from the cell envelope.

Around the outside of the cell membrane is the bacterial cell wall. Bacterial cell walls are made of peptidoglycan (also called murein), which is made from polysaccharide chains cross-linked by unusual peptides containing D-amino acids. Bacterial cell walls are different from the cell walls of plants and fungi which are made of cellulose and chitin, respectively. The cell wall of bacteria is also distinct from that of Archaea, which do not contain peptidoglycan. The cell wall is essential to the survival of many bacteria, although L-form bacteria can be produced in the laboratory that lack a cell wall. The antibiotic penicillin is able to kill bacteria by preventing the cross-linking of peptidoglycan and this causes the cell wall to weaken and lyse. The lysozyme enzyme can also damage bacterial cell walls.

There are broadly speaking two different types of cell wall in bacteria, called Gram-positive and Gram-negative. The names originate from the reaction of cells to the Gram stain, a test long-employed for the classification of bacterial species.

Gram-positive bacteria possess a thick cell wall containing many layers of peptidoglycan and teichoic acids. In contrast, Gram-negative bacteria have a relatively thin cell wall consisting of a few layers of peptidoglycan surrounded by a second lipid membrane containing lipopolysaccharides and lipoproteins. Most bacteria have the Gram-negative cell wall and only the Firmicutes and Actinobacteria (previously known as the low G+C and high G+C Gram-positive bacteria, respectively) have the alternative Gram-positive arrangement. These differences in structure can produce differences in antibiotic susceptibility, for instance vancomycin can kill only Gram-positive bacteria and is ineffective against Gram-negative pathogens, such as *Haemophilus influenzae* or *Pseudomonas aeruginosa*.

Archaeal Cell Walls

Although not truly unique, the cell walls of Archaea are unusual. Whereas peptidoglycan is a standard component of all bacterial cell walls, all archaeal cell walls lack peptidoglycan, with the exception of one group of methanogens. In that group, the peptidoglycan is a modified form very different from the kind found in bacteria. There are four types of cell wall currently known among the Archaea.

One type of archaeal cell wall is that composed of pseudopeptidoglycan (also called pseudomurein). This type of wall is found in some methanogens, such as *Methanobacterium* and *Methanothermus*. While the overall structure of archaeal *pseudo*peptidoglycan superficially resembles that of bacterial peptidoglycan, there are a number of significant chemical differences. Like the peptidoglycan found in bacterial cell walls, pseudopeptidoglycan consists of polymer chains of glycan cross-linked by short peptide connections. However, unlike peptidoglycan, the sugar N-acetylmuramic acid is replaced by N-acetyltalosaminuronic acid, and the two sugars are bonded with a β,1-3 glycosidic linkage instead of β,1-4. Additionally, the cross-linking peptides are L-amino acids rather than D-amino acids as they are in bacteria.

A second type of archaeal cell wall is found in *Methanosarcina* and *Halococcus*. This type of cell wall is composed entirely of a thick layer of polysaccharides, which may be sulfated in the case of *Halococcus*. Structure in this type of wall is complex and as yet is not fully investigated.

A third type of wall among the Archaea consists of glycoprotein, and occurs in the hyperthermophiles, *Halobacterium*, and some methanogens. In *Halobacterium*, the proteins in the wall have a high content of acidic amino acids, giving the wall an overall negative charge. The result is an unstable structure that is stabilized by the presence of large quantities of positive sodium ions that neutralize the charge. Consequently, *Halobacterium* thrives only under conditions with high salinity.

In other Archaea, such as *Methanomicrobium* and *Desulfurococcus*, the wall may be composed only of surface-layer proteins, known as an *S-layer*. S-layers are common in bacteria, where they serve as either the sole cell-wall component or an outer layer in conjunction with polysaccharides. Most Archaea are Gram-negative, though at least one Gram-positive member is known.

Other Cell Coverings

Many protists and bacteria produce other cell surface structures apart from cell walls, external (extracellular matrix) or internal. Many algae have a sheath or envelope of mucilage outside the cell made of exopolysaccharides. Diatoms build a frustule from silica extracted from the surrounding water; radiolarians, foraminiferans, testate amoebae and silicoflagellates also produce a skeleton from minerals, called test in some groups. Many green algae, such as *Halimeda* and the Dasycladales, and some red algae, the Corallinales, encase their cells in a secreted skeleton of calcium carbonate. In each case, the wall is rigid and essentially inorganic. It is the non-living component of cell. Some golden algae, ciliates and choanoflagellates produces a shell-like protective outer covering called lorica. Some dinoflagellates have a theca of cellulose plates, and coccolithophorids have coccoliths.

An extracellular matrix is also present in metazoans. Its composition varies between cells, but collagens are the most abundant protein in the ECM.

Centriole

In cell biology a centriole is a cylindrical cell structure composed mainly of a protein called tubulin that is found in most eukaryotic cells. An associated pair of centrioles, surrounded by a shapeless mass of dense material, called the pericentriolar material, or PCM, makes up a compound structure called a centrosome.

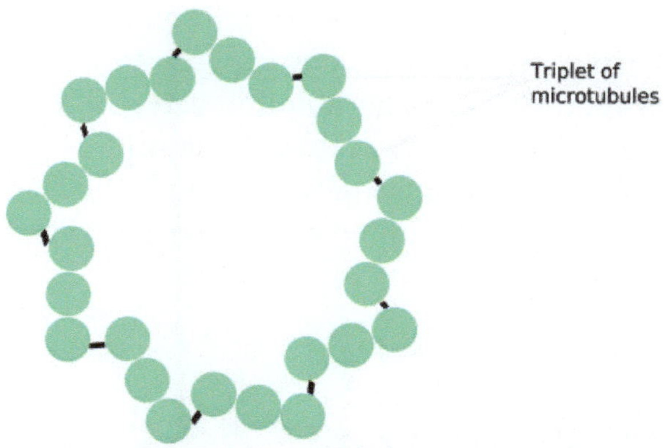

Cross section of centriole - Schema

Centrioles are present in the cells of most eukaryotes, for example those of animals. However, they are absent from conifers (pinophyta), flowering plants (angiosperms) and most fungi, and are only present in the male gametes of charophytes, bryophytes, seedless vascular plants, cycads, and ginkgo.

Most centrioles are made up of nine sets of microtubule triplets, arranged in a cylinder. Deviations from this structure include crabs and *Drosophila melanogaster* embryos, with nine doublets, and *Caenorhabditis elegans* sperm cells and early embryos, with nine singlets.

Edouard van Beneden and Theodor Boveri made the first observation and identification of centrioles in 1883 and 1888 respectively, while the pattern of centriole duplication was first worked out independently by Etienne de Harven and Joseph G. Gall c. 1950 The main function of centrioles is to produce aster and spindle during cell division.

Cell Division

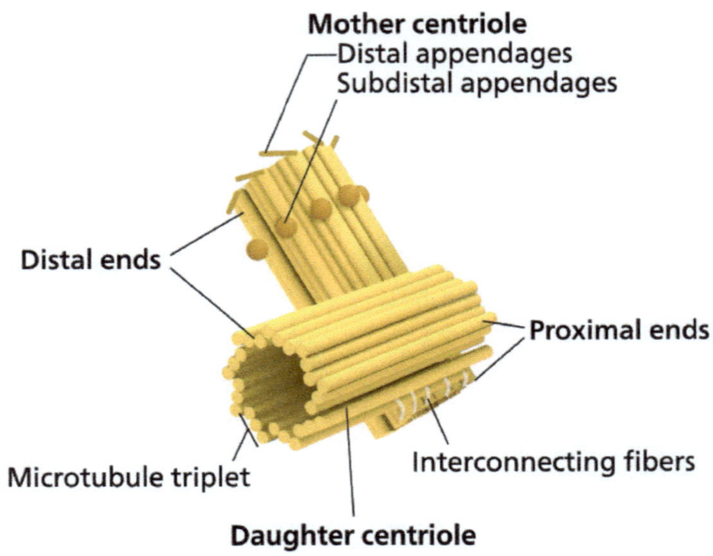

A mother and daughter centriole, attached orthogonally

Centrioles are involved in the organization of the mitotic spindle and in the completion of cytokinesis. Centrioles were previously thought to be required for the formation of a mitotic spindle in animal cells. However, more recent experiments have demonstrated that cells whose centrioles have been removed via laser ablation can still progress through the G_1 stage of interphase before centrioles can be synthesized later in a de novo fashion. Additionally, mutant flies lacking centrioles develop normally, although the adult flies' cells lack flagella and cilia and as a result, they die shortly after birth. The centrioles can self replicate during cell division.

Cellular Organization

Centrioles are a very important part of centrosomes, which are involved in organizing microtubules in the cytoplasm. The position of the centriole determines the position of the nucleus and plays a crucial role in the spatial arrangement of the cell.

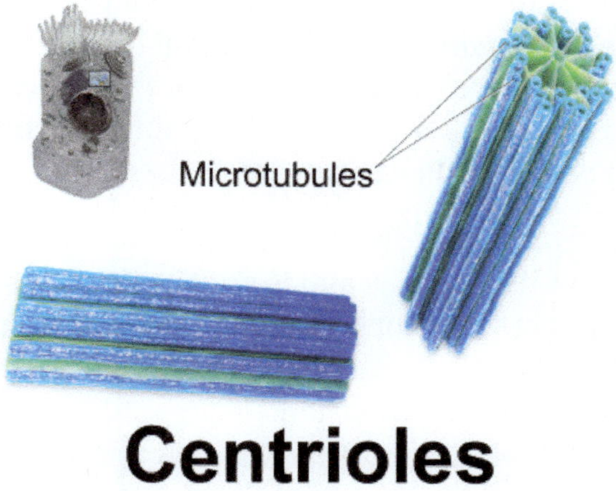

3D Rendering of Centrioles

Ciliogenesis

In organisms with flagella and cilia, the position of these organelles is determined by the mother centriole, which becomes the basal body. An inability of cells to use centrioles to make functional cilia and flagella has been linked to a number of genetic and developmental diseases. In particular, the inability of centrioles to properly migrate prior to ciliary assembly has recently been linked to Meckel-Gruber syndrome.

Animal Development

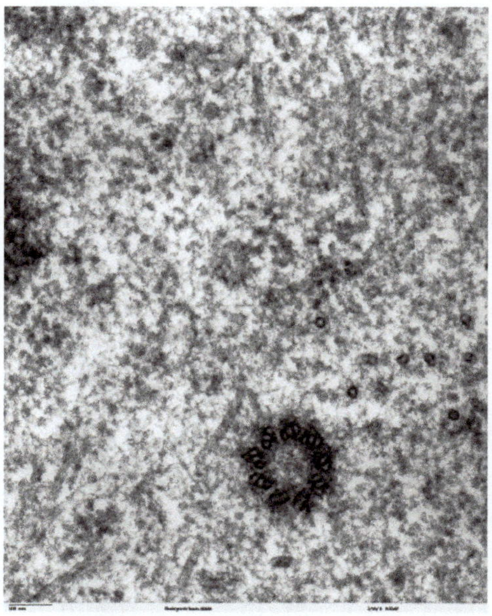

Electron micrograph of a centriole from a mouse embryo.

Proper orientation of cilia via centriole positioning toward the posterior of embryonic node cells is critical for establishing left–right asymmetry during mammalian development.

Centriole Duplication

Before DNA replication, cells contain two centrioles. The older of the two centrioles is termed the *mother centriole*, the other the *daughter*. During the cell division cycle, a new centriole grows from the side of each mother centriole. After duplication, the two centriole pairs will remain attached to each other orthogonally until mitosis. At that point the mother and daughter centrioles separate dependently on an enzyme called separase.

The two centrioles in the centrosome are tied to one another. The mother centriole has radiating appendages at the distal end of its long axis and is attached to its daughter at the proximal end. Each daughter cell formed after cell division will inherit one of these pairs. Centrioles start duplicating when DNA replicates.

Origin

The last common ancestor of all eukaryotes was a ciliated cell with centrioles. Some lineages of eukaryotes, such as land plants, do not have centrioles except in their motile male gametes. Centrioles are completely absent from all cells of conifers and flowering plants, which do not have ciliate or flagellate gametes. It is unclear if the last common ancestor had one or two cilia. Important genes required for centriole growth, like centrins, are only found in eukaryotes and not in bacteria or archaeans.

Etymology and Pronunciation

The word *centriole* combining forms of *centri-* and *-ole*, yielding "little cen-tral part", which describes a centriole's typical location near the center of the cell.

Cytoskeleton

A cytoskeleton is present in all cells of all domains of life (archaea, bacteria, eukaryotes). It is a complex network of interlinking filaments and tubules that extend throughout the cytoplasm, from the nucleus to the plasma membrane The cytoskeletal systems of different organisms are composed of similar proteins. In eukaryotes, the cytoskeletal matrix is a dynamic structure composed of three main proteins, which are capable of rapid growth or disassembly dependent on the cell's requirements at a certain period of time.

The structure, function and dynamic behaviour of the cytoskeleton can be very different, depending on organism and cell type. Even within one cell the cytoskeleton can change through association with other proteins and the previous history of the network.

There is a multitude of functions that the cytoskeleton can perform. Primarily, it gives the cell shape and mechanical resistance to deformation, and through association with extracellular connective tissue and other cells it stabilizes entire tissues. The cytoskeleton can also actively contract, thereby deforming the cell and the cell's environment and allowing cells to migrate. Moreover, it is involved in many cell signaling pathways, in the uptake of extracellular material (endocytosis), segregates chromosomes during cellular division, is involved in cytokinesis (the division of a mother cell into two daughter cells), provides a scaffold to organize the contents of the cell in space and for intracellular transport (for example, the movement of vesicles and organelles within the cell); and can be a template for the construction of a cell wall. Furthermore, it forms specialized structures, such as flagella, cilia, lamellipodia and podosomes.

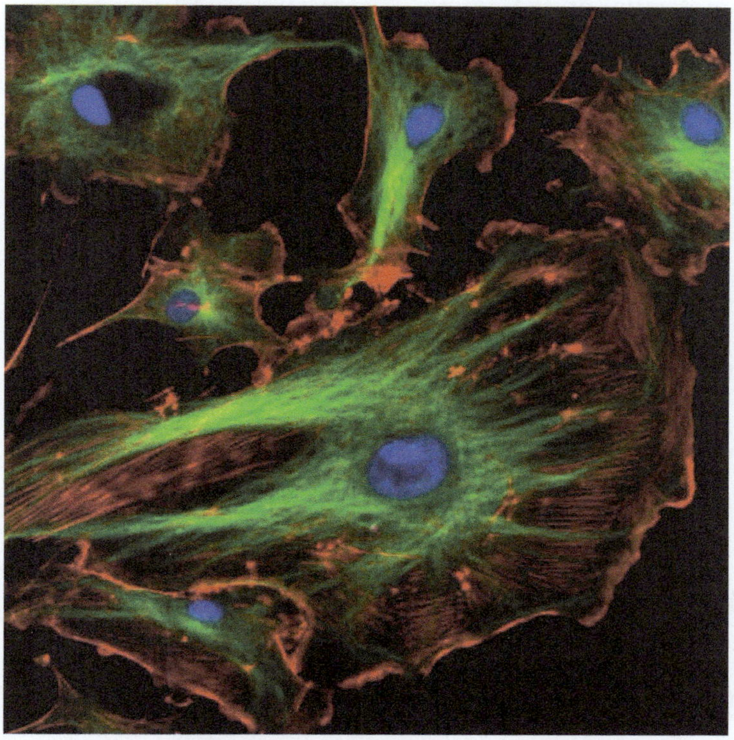

The eukaryotic cytoskeleton. Actin filaments are shown in red, and microtubules composed of beta tubulin are in green.

A large-scale example of an action performed by the cytoskeleton is muscle contraction. During contraction of a muscle, within each muscle cell, myosin molecular motors collectively exert forces on parallel actin filaments. This action contracts the muscle cell, and through the synchronous process in many muscle cells, the entire muscle.

History

In 1903, Nikolai K. Koltsov proposed that the shape of cells was determined by a network of tubules that he termed the cytoskeleton. The concept of a protein mosaic that dynamically coordinated cytoplasmic biochemistry was proposed by Rudolph Peters in 1929 while the term (*cytosquelette*, in French) was first introduced by French embryologist Paul Wintrebert in 1931.

Eukaryotic Cytoskeleton

Eukaryotic cells contain three main kinds of cytoskeletal filaments: microfilaments, microtubules, and intermediate filaments. Each cytoskeletal filament has a shape and intracellular distribution. Additionally, the filaments are formed by polymerization of different types of sub-units. The microfilament consist of the polymers of the protein actin which has a diameter of 7 nm. The microtubules are made up of the protein called tubulin which has a diameter of 25 nm. Intermediate filaments are made up of various proteins which varies depending on the cell type. These type of filament normally have diameters ranging from 8-12 nm. The cytoskeleton provides the cell with structure and shape, and by excluding macromolecules from some of the cytosol, it adds to the level of macromolecular crowding in this compartment. Cytoskeletal elements interact extensively and intimately with cellular membranes. A number of small molecule cytoskeletal drugs have been discovered that interact with actin and microtubules. These compounds have proven useful in studying the cytoskeleton and several have clinical applications.

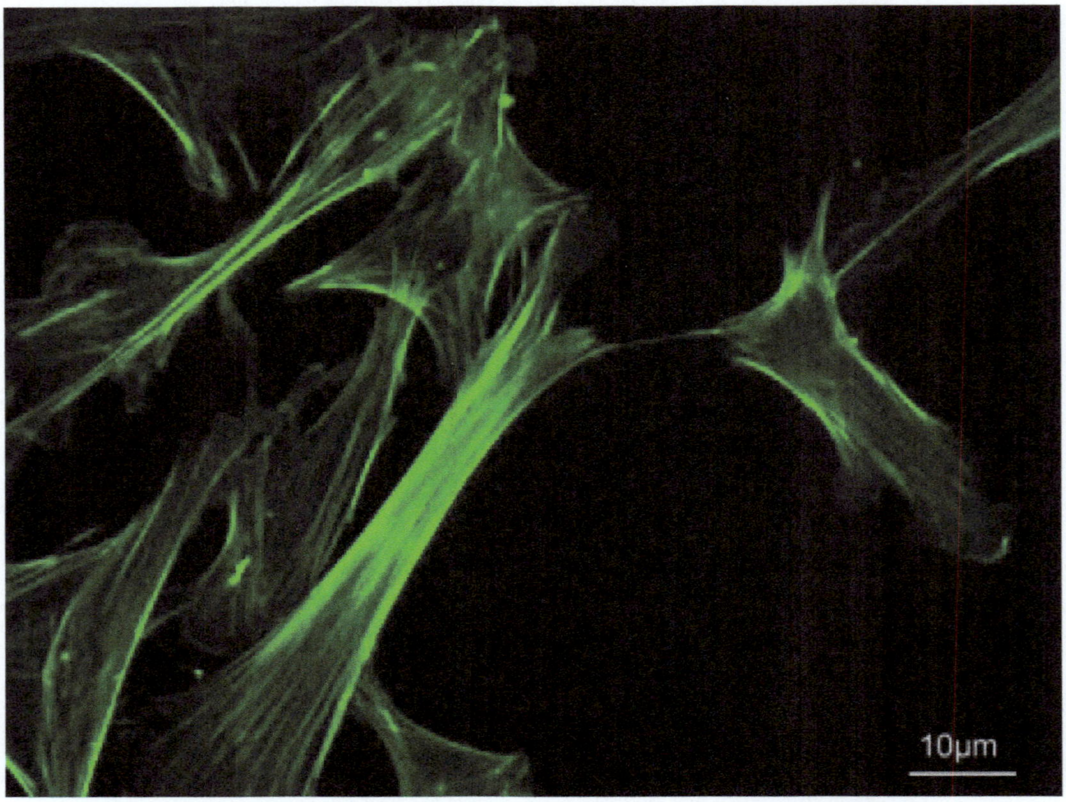

Actin cytoskeleton of mouse embryo fibroblasts, stained with phalloidin

Microfilaments (Actin Filaments)

Microfilaments are composed of linear polymers of G-actin proteins, and generate force when the growing (plus) end of the filament pushes against a barrier, such as the cell membrane. They also act as tracks for the movement of myosin molecules that attach to the microfilament and "walk" along them. Myosin motoring along F-actin filaments generates contractile forces in so-called actomyosin fibers, both in muscle as well as most non-muscle cell types. Actin structures are controlled by the Rho family of small GTP-binding proteins such as Rho itself for contractile acto-myosin filaments ("stress fibers"), Rac for lamellipodia and Cdc42 for filopodia.

Intermediate Filaments

Intermediate filaments are a part of the cytoskeleton of all animals. These filaments, averaging 10 nanometers in diameter, are more stable (strongly bound) than actin filaments, and heterogeneous constituents of the cytoskeleton. Like actin filaments, they function in the maintenance of cell-shape by bearing tension (microtubules, by contrast, resist compression but can also bear tension during mitosis and during the positioning of the centrosome). Intermediate filaments organize the internal tridimensional structure of the cell, anchoring organelles and serving as structural components of the nuclear lamina. They also participate in some cell-cell and cell-matrix junctions. Nuclear lamina exist in all animals and all tissues. Some animals like the fruit fly do not have any cytoplasmic intermediate filaments. In those animals that express cytoplasmic intermediate filaments, these are tissue specific.

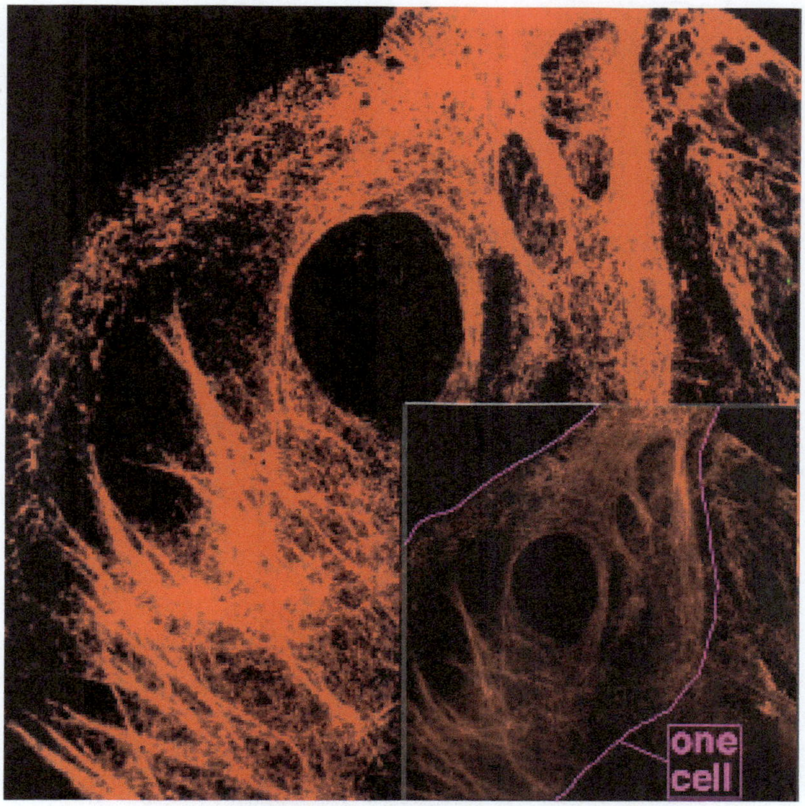

Microscopy of keratin filaments inside cells

Different intermediate filaments are:

- made of vimentins. Vimentin intermediate filaments are in general present in mesenchymal cells.
- made of keratin. Keratin is present in general in epithelial cells.
- neurofilaments of neural cells.
- made of lamin, giving structural support to the nuclear envelope.
- made of desmin, play an important role in structural and mechanical support of muscle cells.

Microtubules

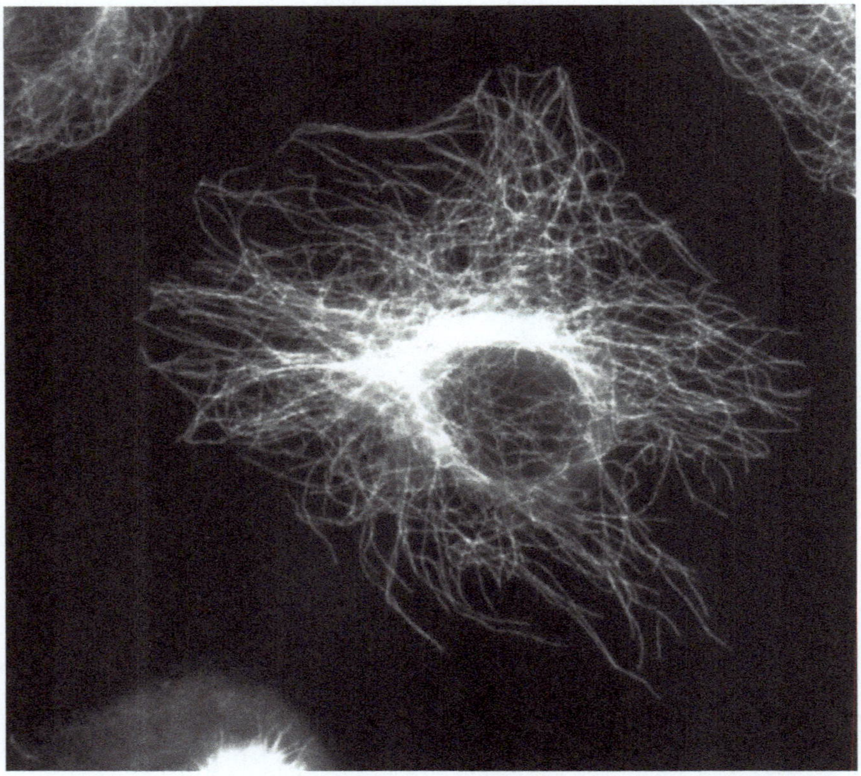

Microtubules in a gel-fixated cell

Microtubules are hollow cylinders about 23 nm in diameter (lumen = approximately 15 nm in diameter), most commonly comprising 13 protofilaments that, in turn, are polymers of alpha and beta tubulin. They have a very dynamic behavior, binding GTP for polymerization. They are commonly organized by the centrosome.

In nine triplet sets (star-shaped), they form the centrioles, and in nine doublets oriented about two additional microtubules (wheel-shaped), they form cilia and flagella. The latter formation is commonly referred to as a "9+2" arrangement, wherein each doublet is connected to another by the protein dynein. As both flagella and cilia are structural components of the cell, and are maintained by microtubules, they can be considered part of the cytoskeleton.

They play key roles in:

- intracellular transport (associated with dyneins and kinesins, they transport organelles like mitochondria or vesicles).
- the axoneme of cilia and flagella.
- the mitotic spindle.
- synthesis of the cell wall in plants.

Comparison

Cytoskeleton type	Diameter (nm)	Structure	Subunit examples
Microfilaments	6	double helix	actin
Intermediate filaments	10	two anti-parallel helices/dimers, forming tetramers	• vimentin (mesenchyme) • glial fibrillary acidic protein (glial cells) • neurofilament proteins (neuronal processes) • keratins (epithelial cells) • nuclear lamins
Microtubules	23	protofilaments, in turn consisting of tubulin subunits in complex with stathmin	α- and β-tubulin

Septins

Septins are a group of the highly conserved GTP binding proteins found in eukaryotes. Different septins form protein complexes with each other. These can assemble to filaments and rings. Therefore, septins can be considered part of the cytoskeleton. The function of septins in cells include serving as a localized attachment site for other proteins, and preventing the diffusion of certain molecules from one cell compartment to another. In yeast cells, they build scaffolding to provide structural support during cell division and compartmentalize parts of the cell. Recent research in human cells suggests that septins build cages around bacterial pathogens, immobilizing the harmful microbes and preventing them from invading other cells.

Spectrin

Spectrin is a cytoskeletal protein that lines the intracellular side of the plasma membrane in eukaryotic cells. Spectrin forms pentagonal or hexagonal arrangements, forming a scaffolding and playing an important role in maintenance of plasma membrane integrity and cytoskeletal structure.

Yeast cytoskeleton

In budding yeast (an important model organism), actin forms cortical patches, actin cables, and a cytokinetic ring and the cap. Cortical patches are discrete actin bodies on the membrane and are important for endocytosis, especially the recycling of glucan synthase which is important for cell wall synthesis. Actin cables are bundles of actin filaments and are involved in the transport of vesicles towards the cap (which contains a number of different proteins to polarize cell growth) and in the positioning of mitochondria. The cytokinetic ring forms and constricts around the site of cell division.

Prokaryotic cytoskeleton

The cytoskeleton was once thought to be a feature only of eukaryotic cells, but homologues to all the major proteins of the eukaryotic cytoskeleton have been found in prokaryotes. Although the evolutionary relationships are so distant that they are not obvious from protein sequence comparisons alone, the similarity of their three-dimensional structures and similar functions in maintaining cell shape and polarity provides strong evidence that the eukaryotic and prokaryotic cytoskeletons are truly homologous. However, some structures in the bacterial cytoskeleton may have yet to be identified.

FtsZ

FtsZ was the first protein of the prokaryotic cytoskeleton to be identified. Like tubulin, FtsZ forms filaments in the presence of guanosine triphosphate (GTP), but these filaments do not group into tubules. During cell division, FtsZ is the first protein to move to the division site, and is essential for recruiting other proteins that synthesize the new cell wall between the dividing cells.

MreB and ParM

Prokaryotic actin-like proteins, such as MreB, are involved in the maintenance of cell shape. All non-spherical bacteria have genes encoding actin-like proteins, and these proteins form a helical network beneath the cell membrane that guides the proteins involved in cell wall biosynthesis.

Some plasmids encode a partitioning system that involves an actin-like protein ParM. Filaments of ParM exhibit dynamic instability, and may partition plasmid DNA into the dividing daughter cells by a mechanism analogous to that used by microtubules during eukaryotic mitosis.

Crescentin

The bacterium *Caulobacter crescentus* contains a third protein, crescentin, that is related to the intermediate filaments of eukaryotic cells. Crescentin is also involved in maintaining cell shape, such as helical and vibrioid forms of bacteria, but the mechanism by which it does this is currently unclear.

Common Features and Differences Between Prokaryotes and Eukaryotes

By definition, the cytoskeleton is composed of proteins that can form longitudinal arrays (fibres) in all organisms. These filament forming proteins have been classified into 4 classes. Tubulin-like, actin-like, Walker A cytoskeletal ATPases (WACA-proteins), and intermediate filaments.

Tubulin-like proteins are tubulin in eukaryotes and FtsZ, TubZ, RepX in prokaryotes. Actin-like proteins are actin in eukaryotes and MreB, FtsA in prokaryotes. An example of a WACA-proteins, which are mostly found in prokaryotes, is MinD. Examples for intermediate filaments, which have almost exclusively been found in animals (i.e. eukaryotes) are the lamins, keratins, vimentin, neurofilaments, desmin.

Although tubulin-like proteins share some amino acid sequence similarity, their similarity in protein-fold and the similarity in the GTP binding site is more striking. The same holds true for the actin-like proteins and their structure and ATP binding domain.

Cytoskeletal proteins are usually associated with cell shape, DNA segregation and cell division in prokaryotes and eukaryotes. Which proteins fulfill which task is very different. For example, DNA segregation in all eukaryotes happens through use of tubulin, but in prokaryotes either WACA proteins, actin-like or tubulin-like proteins can be used. Cell division is mediated in eukaryotes by actin, but in prokaryotes usually by tubulin-like (often FtsZ-ring) proteins and sometimes (Crenarchaeota) ESCRT-III, which in eukaryotes still has a role in the last step of division.

Cytoplasm

In cell biology, the cytoplasm is the material or protoplasm within a living cell, excluding the cell nucleus. It comprises cytosol (the gel-like substance enclosed within the cell membrane) and the organelles – the cell's internal sub-structures. All of the contents of the cells of prokaryote organisms (such as bacteria, which lack a cell nucleus) are contained within the cytoplasm. Within the cells of eukaryote organisms the contents of the cell nucleus are separated from the cytoplasm, and are then called the nucleoplasm. The cytoplasm is about 80% water and usually colorless.

It is within the cytoplasm that most cellular activities occur, such as many metabolic pathways including glycolysis, and processes such as cell division. The concentrated inner area is called the endoplasm and the outer layer is called the cell cortex or the ectoplasm.

Movement of calcium ions in and out of the cytoplasm is a signaling activity for metabolic processes.

In plants, movement of the cytoplasm around vacuoles is known as cytoplasmic streaming.

History

The term was introduced by Rudolf von Kölliker in 1863, originally as a synonym for protoplasm, but later it has come to mean the cell substance and organelles outside the nucleus.

There has been certain disagreement on the definition of cytoplasm, as some authors prefer to exclude from it some organelles, specially the vacuoles and sometimes the plastids.

Physical Nature of Cytoplasm

The physical properties of the cytoplasm have been contested in recent years.It remains uncertain how the varied components of the cytoplasm interact to allow movement of particles and organelles while maintaining the cell's structure. The flow of cytoplasmic components plays an important role in many cellular functions which are dependent on the permeability of the cytoplasm. An obvious example of such function is cell signalling, a process which is dependent on the manner in which signaling molecules are allowed to diffuse across the cell. While small signaling molecules like calcium ions are able to diffuse with ease, larger molecules and subcellular structures often require aid in moving through the cytoplasm. The irregular dynamics of such particles have given rise to various theories on the nature of the cytoplasm.

Cytoplasm as a Sol-Gel

There has long been evidence that the cytoplasm behaves like a sol-gel. It is thought that the component molecules and structures of the cytoplasm behave at times like a disordered colloidal solution (sol) and at other times like an integrated network, forming a solid mass (gel). This theory thus proposes that the cytoplasm exists in distinct fluid and solid phases depending on the level of interaction between cytoplasmic components, which may explain the differential dynamics of different particles observed moving through the cytoplasm.

Cytoplasm as a Glass

Recently it has been proposed that the cytoplasm behaves like a glass-forming liquid approaching the glass transition. In this theory, the greater the concentration of cytoplasmic components, the less the cytoplasm behaves like a liquid and the more it behaves as a solid glass, freezing larger cytoplasmic components in place (it is thought that the cell's metabolic activity is able to fluidize the cytoplasm to allow the movement of such larger cytoplasmic components). A cell's ability to vitrify in the absence of metabolic activity, as in dormant periods, may be beneficial as a defence strategy. A solid glass cytoplasm would freeze subcellular structures in place, preventing damage, while allowing the transmission of very small proteins and metabolites, helping to kickstart growth upon the cell's revival from dormancy.

Other Perspectives

There has been research examining the motion of cytoplasmic particles independent of the nature of the cytoplasm. In such an alternative approach, the aggregate random forces within the cell caused by motor proteins explain the non-Brownian motion of cytoplasmic constituents.

Constituents

The three major elements of the cytoplasm are the cytosol, organelles and inclusions.

Cytosol

The cytosol is the portion of the cytoplasm not contained within membrane-bound organelles. Cytosol makes up about 70% of the cell volume and is a complex mixture of cytoskeleton filaments, dissolved molecules, and water. The cytosol's filaments include the protein filaments such as actin filaments and microtubules that make up the cytoskeleton, as well as soluble proteins and small structures such as ribosomes, proteasomes, and the mysterious vault complexes. The inner, granular and more fluid portion of the cytoplasm is referred to as endoplasm.

Due to this network of fibres and high concentrations of dissolved macromolecules, such as proteins, an effect called macromolecular crowding occurs and the cytosol does not act as an ideal solution. This crowding effect alters how the components of the cytosol interact with each other.

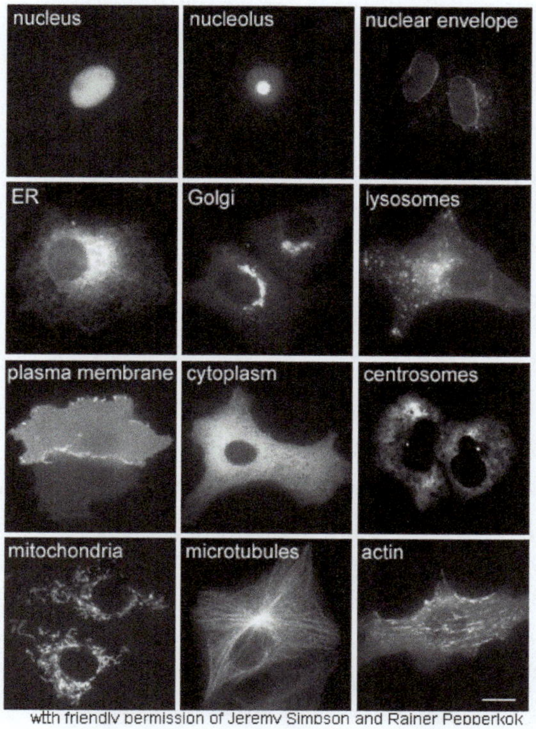

with friendly permission of Jeremy Simpson and Rainer Pepperkok

Proteins in different cellular compartments and structures tagged with green fluorescent protein

Organelles

Organelles (literally "little organs"), are usually membrane-bound structures inside the cell that have specific functions. Some major organelles that are suspended in the cytosol are the mitochondria, the endoplasmic reticulum, the Golgi apparatus, vacuoles, lysosomes, and in plant cells chloroplasts.

Cytoplasmic inclusions

The inclusions are small particles of insoluble substances suspended in the cytosol. A huge range of inclusions exist in different cell types, and range from crystals of calcium oxalate or silicon dioxide in plants, to granules of energy-storage materials such as starch, glycogen, or polyhydroxybutyrate. A particularly widespread example are lipid droplets, which are spherical droplets composed of lipids and proteins that are used in both prokaryotes and eukaryotes as a way of storing lipids such as fatty acids and sterols. Lipid droplets make up much of the volume of adipocytes, which are specialized lipid-storage cells, but they are also found in a range of other cell types.

Controversy and Research

The cytoplasm, mitochondria and most organelles are contributions to the cell from the maternal gamete. Contrary to the older information that disregards any notion of the cytoplasm being active, new research has shown it to be in control of movement and flow of nutrients in and out of the cell by viscoplastic behavior and a measure of the reciprocal rate of bond breakage within the cytoplasmic network.

The material properties of the cytoplasm remain an ongoing investigation. Recent measurements using force spectrum microscopy reveal that the cytoplasm can be likened to an elastic solid, rather than a viscoelastic fluid.

Plasmid

A plasmid is a small DNA molecule within a cell that is physically separated from a chromosomal DNA and can replicate independently. They are most commonly found in bacteria as small circular, double-stranded DNA molecules; however, plasmids are sometimes present in archaea and eukaryotic organisms. In nature, plasmids often carry genes that may benefit the survival of the organism, for example antibiotic resistance. While the chromosomes are big and contain all the essential genetic information for living under normal conditions, plasmids usually are very small and contain only additional genes that may be useful to the organism under certain situations or particular conditions. Artificial plasmids are widely used as vectors in molecular cloning, serving to drive the replication of recombinant DNA sequences within host organisms.

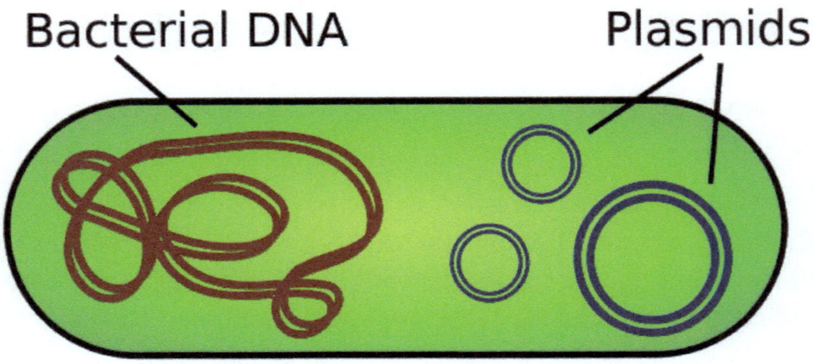

Illustration of a bacterium showing chromosomal DNA and plasmids. Not to scale.

Plasmids are considered *replicons*, a unit of DNA capable of replicating autonomously within a suitable host. However, plasmids, like viruses, are not generally classified as life. Plasmids can be transmitted from one bacterium to another (even of another species) via three main mechanisms: transformation, transduction, and conjugation. This host-to-host transfer of genetic material is called horizontal gene transfer, and plasmids can be considered part of the mobilome. Unlike viruses (which encase their genetic material in a protective protein coat called a capsid), plasmids are "naked" DNA and do not encode genes necessary to encase the genetic material for transfer to a new host. However, some classes of plasmids encode the conjugative "sex" pilus necessary for their own transfer. The size of the plasmid varies from 1 to over 200 kbp, and the number of identical plasmids in a single cell can range anywhere from one to thousands under some circumstances.

The relationship between microbes and plasmid DNA is neither parasitic nor mutualistic, because each implies the presence of an independent species living in a detrimental or commensal state with the host organism. Rather, plasmids provide a mechanism for horizontal gene transfer within a population of microbes and typically provide a selective advantage under a given environmental state. Plasmids may carry genes that provide resistance to naturally occurring antibiotics in a competitive environmental niche, or the proteins produced may act as toxins under similar circumstances, or allow the organism to utilize particular organic compounds that would be advantageous when nutrients are scarce.

Properties and Characteristics

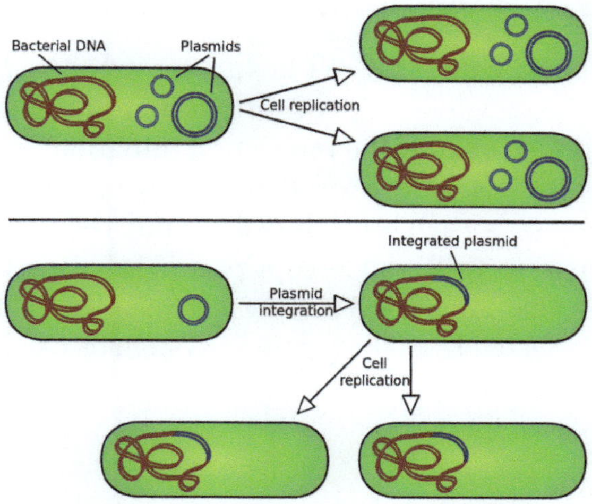

There are two types of plasmid integration into a host bacteria: Non-integrating plasmids replicate as with the top instance, whereas episomes, the lower example, can integrate into the host chromosome.

The American molecular biologist Joshua Lederberg first introduced the term *plasmid* in 1952 - originally to describe any bacterial genetic material that exists in an extrachromosomal state for at least part of its replication cycle. Later in 1968, it was decided that the term plasmid should be adopted as the term for extrachromosomal genetic element, and to distinguish it from viruses, the definition was narrowed to genetic elements that exist exclusively or predominantly outside of the chromosome and can replicate autonomously.

In order for plasmids to replicate independently within a cell, they must possess a stretch of DNA that can act as an origin of replication. The self-replicating unit, in this case the plasmid, is called a replicon. A typical bacterial replicon may consist of a number of elements, such as the gene for plasmid-specific replication initiation protein (Rep), repeating units called iterons, DnaA boxes, and an adjacent AT-rich region. Smaller plasmids make use of the host replicative enzymes to make copies of themselves, while larger plasmids may carry genes specific for the replication of those plasmids. A few types of plasmids can also insert into the host chromosome, and these integrative plasmids are sometimes referred to as episomes in prokaryotes.

Plasmids almost always carry at least one gene. Many of the genes carried by a plasmid are beneficial for the host cells, for example: enabling the host cell to survive in an environment that would otherwise be lethal or restrictive for growth. Some of these genes encode traits for antibiotic resistance or resistance to heavy metal, while others may produce virulence factors that enable a bacterium to colonize a host and overcome its defences, or have specific metabolic functions that allow the bacterium to utilize a particular nutrient, including the ability to degrade recalcitrant or toxic organic compounds. Plasmids can also provide bacteria with the ability to fix nitrogen. Some plasmids, however, have no observable effect on the phenotype of the host cell or its benefit to the host cells cannot be determined, and these plasmids are called cryptic plasmids.

Natural occurring plasmids vary greatly in their physical properties. Their size can range from very small mini-plasmids of less than a 1 kilobase pairs (Kbp), to very large megaplasmids of several megabase pairs (Mbp). At the upper end, little can differentiate between a megaplasmid and a minichromosome. Plasmids are generally circular, however examples of linear plasmids are also known. These linear plasmids require specialized mechanisms to replicate their ends.

Plasmids may be present in an individual cell in varying number, ranging from one to several hundreds. The normal number of copies of plasmid that may be found in a single cell is called the copy number, and is determined by how the replication initiation is regulated and the size of the molecule. Larger plasmids tend to have lower copy numbers. Low-copy-number plasmids that exist only as one or a few copies in each bacterium are, upon cell division, in danger of being lost in one of the segregating bacteria. Such single-copy plasmids have systems that attempt to actively distribute a copy to both daughter cells. These systems, which include the parABS system and parMRC system, are often referred to as the partition system or partition function of a plasmid.

Classifications and Types

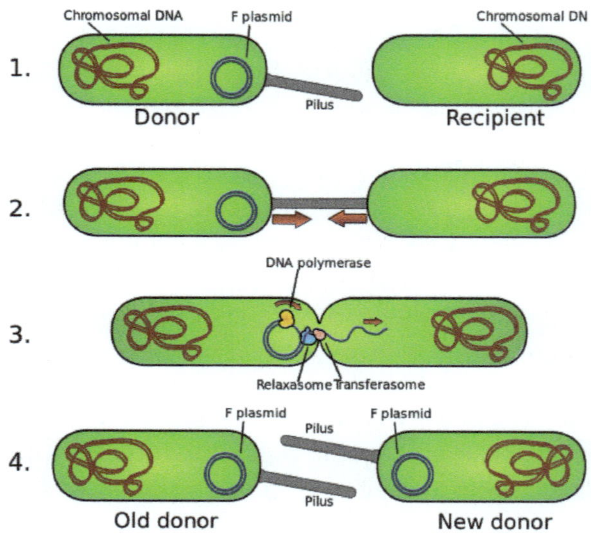

Overview of bacterial conjugation

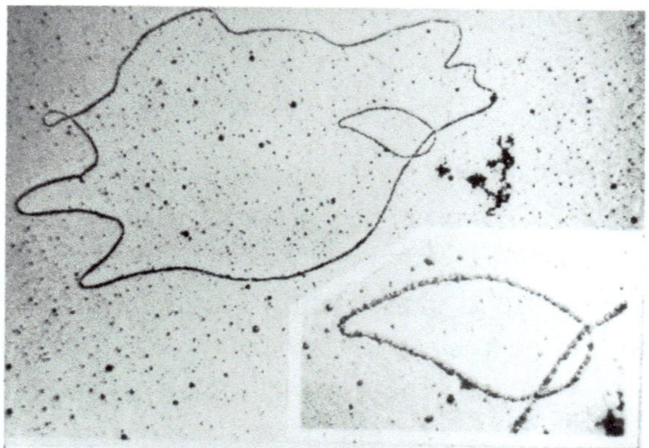

Electron micrograph of a DNA fiber bundle, presumably of a single bacterial chromosome loop.

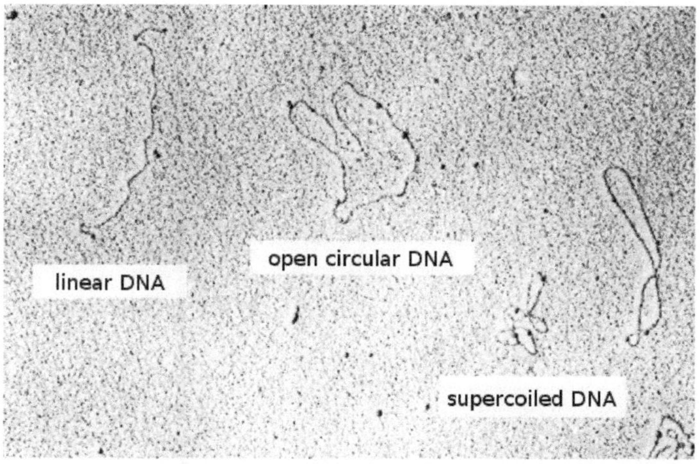

Electron micrograph of a bacterial DNA plasmid (chromosome fragment).

Plasmids may be classified in a number of ways. Plasmids can be broadly classified into conjugative plasmids and non-conjugative plasmids. Conjugative plasmids contain a set of transfer or *tra* genes which promote sexual conjugation between different cells. In the complex process of conjugation, plasmid may be transferred from one bacterium to another via sex pili encoded by some of the *tra* genes. Non-conjugative plasmids are incapable of initiating conjugation, hence they can be transferred only with the assistance of conjugative plasmids. An intermediate class of plasmids are mobilizable, and carry only a subset of the genes required for transfer. They can parasitize a conjugative plasmid, transferring at high frequency only in its presence.

Plasmids can also be classified into incompatibility group. A microbe can harbour different types of plasmids, however, different plasmids can only exist in a single bacterial cell if they are compatible. If two plasmids are not compatible, one or the other will be rapidly lost from the cell. Different plasmids may therefore be assigned to different incompatibility group depending on whether they can coexist together. Incompatible plasmids normally share the same replication or partition mechanisms.

Another way to classify plasmids is by function. There are five main classes:

- Fertility F-plasmids, which contain *tra* genes. They are capable of conjugation and result in the expression of sex pili.
- Resistance (R) plasmids, which contain genes that provide resistance against antibiotics or poisons. Historically known as R-factors, before the nature of plasmids was understood.
- Col plasmids, which contain genes that code for bacteriocins, proteins that can kill other bacteria.
- Degradative plasmids, which enable the digestion of unusual substances, e.g. toluene and salicylic acid.
- Virulence plasmids, which turn the bacterium into a pathogen.

Plasmids can belong to more than one of these functional groups.

Vectors

Artificially constructed plasmids may be used as vectors in genetic engineering. These plasmids serve as important tools in genetics and biotechnology labs, where they are commonly used to clone and amplify (make many copies of) or express particular genes. A wide variety of plasmids are commercially available for such uses. The gene to be replicated is normally inserted into a plasmid that typically contains a number of features for their use. These include a gene that confers resistance to particular antibiotics (ampicillin is most frequently used for bacterial strains), an origin of replication to allow the bacterial cells to replicate the plasmid DNA, and a suitable site for cloning.

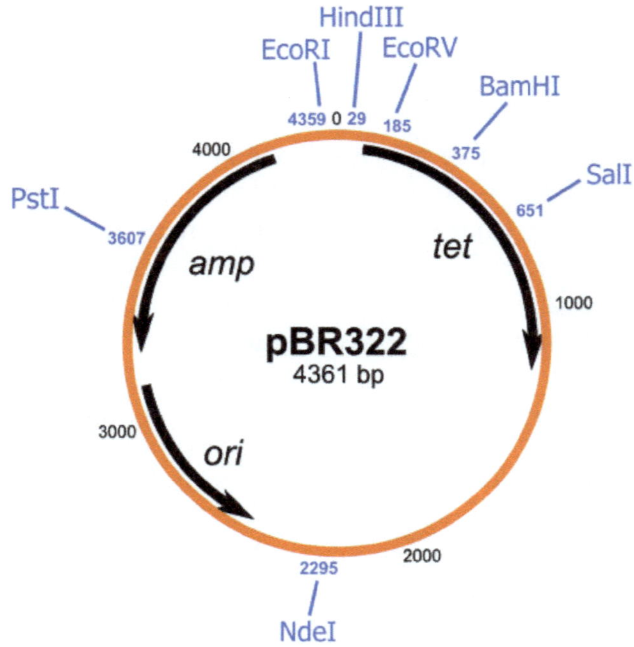

A schematic representation of the pBR322 plasmid, one of the first plasmids to be used widely as a cloning vector. Shown on the plasmid diagram are the genes encoded (*amp* and *tet* for ampicillin and tetracycline resistance respectively), its origin of replication (*ori*), and various restriction sites (indicated in blue).

Cloning

Plasmids are the most-commonly used bacterial cloning vectors. These cloning vectors contain a site that allows DNA fragments to be inserted, for example a multiple cloning site or polylinker which has several commonly used restriction sites to which DNA fragments may be ligated. After the gene of interest is inserted, the plasmids are introduced into bacteria by a process called transformation. These plasmids contain a selectable marker, usually an antibiotic resistance gene, which confer on the bacteria an ability to survive and proliferate in a selective growth medium containing the particular antibiotics. The cells after transformation are exposed to the selective media, and only cells containing the plasmid may survive. In this way, the antibiotics act as a filter to select only the bacteria containing the plasmid DNA. The vector may also contain other marker genes or reporter genes to facilitate selection of plasmid with cloned insert. Bacteria containing the plasmid can then be grown in large amounts, harvested, and the plasmid of interest may then be isolated using various methods of plasmid preparation.

A plasmid cloning vector is typically used to clone DNA fragments of up to 15 kbp. To clone longer lengths of DNA, lambda phage with lysogeny genes deleted, cosmids, bacterial artificial chromosomes, or yeast artificial chromosomes are used.

Protein Production

Another major use of plasmids is to make large amounts of proteins. In this case, researchers grow bacteria containing a plasmid harboring the gene of interest. Just as the bacterium produces proteins to confer its antibiotic resistance, it can also be induced to produce large amounts of proteins from the inserted gene. This is a cheap and easy way of mass-producing the protein the gene codes for, for example, insulin.

Gene Therapy

Plasmid may also be used for gene transfer into human cells as potential treatment in gene therapy so that it may express the protein that is lacking in the cells. Some strategies of gene therapy require the insertion of therapeutic genes at pre-selected chromosomal target sites within the human genome. Plasmid vectors are one of many approaches that could be used for this purpose. Zinc finger nucleases (ZFNs) offer a way to cause a site-specific double-strand break to the DNA genome and cause homologous recombination. Plasmids encoding ZFN could help deliver a therapeutic gene to a specific site so that cell damage, cancer-causing mutations, or an immune response is avoided.

Disease Models

Plasmids were historically used to genetically engineer the embryonic stem cells of rats in order to create rat genetic disease models. The limited efficiency of plasmid-based techniques precluded their use in the creation of more accurate human cell models. However, developments in Adeno-associated virus recombination techniques, and Zinc finger nucleases, have enabled the creation of a new generation of isogenic human disease models.

Episomes

The term episome was proposed by François Jacob and Élie Wollman in 1958 to describe extra-chromosomal genetic material that may replicate autonomously or become integrated into the chromosome. The use of the term, however, has diverged since it was first coined as plasmid became the preferred word for autonomously replicating extrachromosomal DNA as proposed in the symposium in 1968 – it was suggested by some that the use of the term episome be abandoned, although others continued to use the term with a shift in meaning. In prokaryotes, episome is now used by some to refer to plasmid that is capable of integrating into the chromosome. The integrative plasmids may be replicated and stably maintained in a cell through multiple generations, but always at some stage they exist as an independent plasmid molecule.

In eukaryotes, episomes are used to mean non-integrated extrachromosomal closed circular DNA molecule that may be replicated in the nucleus. Viruses are the most common examples of this, such as herpesviruses, adenoviruses, and polyomaviruses, but some are plasmids. Other examples include aberrant chromosomal fragments, such as double minute chromosomes, that can arise during artificial gene amplifications or in pathologic processes (e.g., cancer cell transformation). Episomes in eukaryotes behave similarly to plasmids in prokaryotes in that the DNA is stably maintained and replicated with the host cell. Cytoplasmic viral episomes (as in poxvirus infections) can also occur. Some episomes, such as herpesviruses, replicate in a rolling circle mechanism, similar to bacterial phage viruses. Others replicate through a bidirectional replication mechanism (*Theta type* plasmids). In either case, episomes remain physically separate from host cell chromosomes. Several cancer viruses, including Epstein-Barr virus and Kaposi's sarcoma-associated herpesvirus, are maintained as latent, chromosomally distinct episomes in cancer cells, where the viruses express oncogenes that promote cancer cell proliferation. In cancers, these episomes passively replicate together with host chromosomes when the cell divides. When these viral episomes initiate lytic replication to generate multiple virus particles, they in general activate cellular innate immunity defense mechanisms that kill the host cell.

Plasmid Maintenance

Some plasmids or microbial hosts include an addiction system or postsegregational killing system (PSK), such as the hok/sok (host killing/suppressor of killing) system of plasmid R1 in *Escherichia coli*. This variant produces both a long-lived poison and a short-lived antidote. Several types of plasmid addiction systems (toxin/ antitoxin, metabolism-based, ORT systems) were described in the literature and used in biotechnical (fermentation) or biomedical (vaccine therapy) applications. Daughter cells that retain a copy of the plasmid survive, while a daughter cell that fails to inherit the plasmid dies or suffers a reduced growth-rate because of the lingering poison from the parent cell. Finally, the overall productivity could be enhanced.

In contrast, virtually all biotechnologically used plasmids (such as pUC18, pBR322 and derived vectors) do not contain toxin-antitoxin addiction systems and thus need to be kept under antibiotic pressure to avoid plasmid loss.

Yeast Plasmids

Yeast are organisms that naturally harbour plasmids. Notable plasmids are 2 μm plasmids - small circular plasmids often used for genetic engineering of yeast, and linear pGKL plasmids from *Kluyveromyces lactis*, that are responsible for killer phenotypes.

Other types of plasmids are often related to yeast cloning vectors that include:

- *Yeast integrative plasmid (YIp)*, yeast vectors that rely on integration into the host chromosome for survival and replication, and are usually used when studying the functionality of a solo gene or when the gene is toxic. Also connected with the gene URA3, that codes an enzyme related to the biosynthesis of pyrimidine nucleotides (T, C);
- *Yeast Replicative Plasmid (YRp)*, which transport a sequence of chromosomal DNA that includes an origin of replication. These plasmids are less stable, as they can *get lost* during the budding.

Plasmid DNA Extraction

As alluded to above, plasmids are often used to purify a specific sequence, since they can easily be purified away from the rest of the genome. For their use as vectors, and for molecular cloning, plasmids often need to be isolated.

There are several methods to isolate plasmid DNA from bacteria, the archetypes of which are the miniprep and the maxiprep/bulkprep. The former can be used to quickly find out whether the plasmid is correct in any of several bacterial clones. The yield is a small amount of impure plasmid DNA, which is sufficient for analysis by restriction digest and for some cloning techniques.

In the latter, much larger volumes of bacterial suspension are grown from which a maxi-prep can be performed. In essence, this is a scaled-up miniprep followed by additional purification. This results in relatively large amounts (several hundreds micrograms) of very pure plasmid DNA.

In recent times, many commercial kits have been created to perform plasmid extraction at various scales, purity, and levels of automation. Commercial services can prepare plasmid DNA at quoted prices below $300/mg in milligram quantities and $15/mg in gram quantities (early 2007).

Conformations

Plasmid DNA may appear in one of five conformations, which (for a given size) run at different speeds in a gel during electrophoresis. The conformations are listed below in order of electrophoretic mobility (speed for a given applied voltage) from slowest to fastest:

- *Nicked open-circular* DNA has one strand cut.
- *Relaxed circular* DNA is fully intact with both strands uncut, but has been enzymatically *relaxed* (supercoils removed). This can be modeled by letting a twisted extension cord unwind and relax and then plugging it into itself.
- *Linear* DNA has free ends, either because both strands have been cut or because the DNA was linear *in vivo*. This can be modeled with an electrical extension cord that is not plugged into itself.
- *Supercoiled* (or *covalently closed-circular*) DNA is fully intact with both strands uncut, and with an integral twist, resulting in a compact form. This can be modeled by twisting an extension cord and then plugging it into itself.
- *Supercoiled denatured* DNA is like *supercoiled DNA*, but has unpaired regions that make it slightly less compact; this can result from excessive alkalinity during plasmid preparation.

The rate of migration for small linear fragments is directly proportional to the voltage applied at low voltages. At higher voltages, larger fragments migrate at continuously increasing yet different rates. Thus, the resolution of a gel decreases with increased voltage.

At a specified, low voltage, the migration rate of small linear DNA fragments is a function of their length. Large linear fragments (over 20 kb or so) migrate at a certain fixed rate regardless of length. This is because the molecules 'resperate', with the bulk of the molecule following the leading end through the gel matrix. Restriction digests are frequently used to analyse purified plasmids. These enzymes specifically break the DNA at certain short sequences. The resulting linear fragments form 'bands' after gel electrophoresis. It is possible to purify certain fragments by cutting the bands out of the gel and dissolving the gel to release the DNA fragments.

Because of its tight conformation, supercoiled DNA migrates faster through a gel than linear or open-circular DNA.

Software for bioinformatics and design

The use of plasmids as a technique in molecular biology is supported by bioinformatics software. These programs record the DNA sequence of plasmid vectors, help to predict cut sites of restriction enzymes, and to plan manipulations. Examples of software packages that handle plasmid maps are ApE, Clone Manager, GeneConstructionKit, Geneious, Genome Compiler, LabGenius, Lasergene, MacVector, pDraw32, VectorFriends, Vector NTI, and WebDSV. These software help conduct entire experiments in silico before doing wet experiments .

Stroma

Biology

- Stroma (animal tissue), the connective, functionally supportive framework of a biological cell, tissue, or organ (in contrast, the parenchyma is the functional aspect of a tissue).
    - Stroma of ovary, a soft tissue, well supplied with blood, consisting of spindle-shaped cells with a small amount of connective tissue
    - Stroma of iris, fibres and cells in the iris
    - Stroma of cornea, plates of collagen fibrils in the cornea
    - Lymph node stromal cell, cells which provide a scaffold for other lymph node cells
- Stroma (fluid), the fluid in between grana, where carbohydrate-formation reactions occur in the chloroplasts of plant cells photosynthesizing
- Stroma (mycology), the dense structural tissue that produces fruiting bodies; made of non-vegetative (asexual) hyphae
- Stromal cell, a connective tissue cell of any organ, supports the function of the parenchymal cells of that organ

Places

- Stroma, Scotland, an uninhabited island off the northern coast of Scotland

People

- Freddie Stroma (1987-), a British actor known for playing Cormac McLaggen in *Harry Potter and the Half-Blood Prince*

Other

- Stroma (musical group), a New Zealand chamber music ensemble

Cytosol

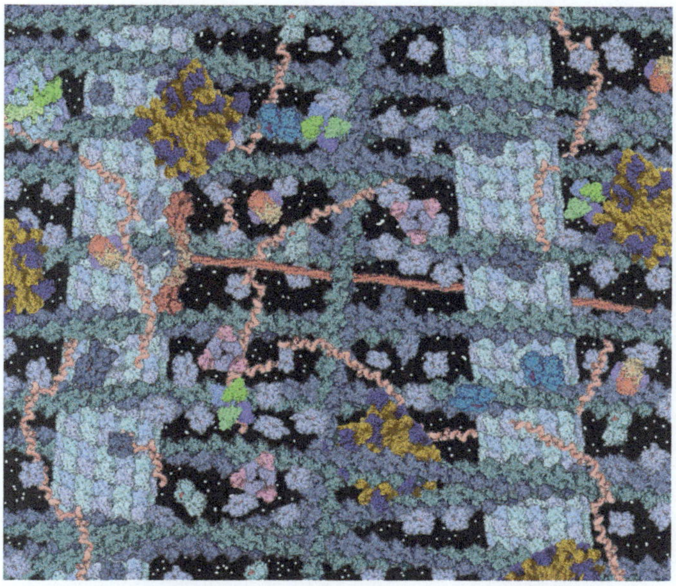

The cytosol is a crowded solution of many different types of molecules that fills much of the volume of cells.

The cytosol or *cytoplasmic matrix* is the liquid found inside cells. It constitutes most of the intracellular fluid (ICF). It is separated into compartments by membranes. For example, the mitochondrial matrix separates the mitochondrion into many compartments.

In the eukaryotic cell, the cytosol is within the cell membrane and is part of the cytoplasm, which also comprises the mitochondria, plastids, and other organelles (but not their internal fluids and structures); the cell nucleus is separate. The cytosol is thus a liquid matrix around the organelles. In prokaryotes, most of the chemical reactions of metabolism take place in the cytosol, while a few take place in membranes or in the periplasmic space. In eukaryotes, while many metabolic pathways still occur in the cytosol, others are contained within organelles.

The cytosol is a complex mixture of substances dissolved in water. Although water forms the large majority of the cytosol, its structure and properties within cells is not well understood. The concentrations of ions such as sodium and potassium are different in the cytosol than in the extracellular fluid; these differences in ion levels are important in processes such as osmoregulation, cell signaling, and the generation of action potentials in excitable cells such as endocrine, nerve and muscle cells. The cytosol also contains large amounts of macromolecules, which can alter how molecules behave, through macromolecular crowding.

Although it was once thought to be a simple solution of molecules, the cytosol has multiple levels of organization. These include concentration gradients of small molecules such as calcium, large complexes of enzymes that act together to carry out metabolic pathways, and protein complexes such as proteasomes and carboxysomes that enclose and separate parts of the cytosol.

Definition

The term "cytosol" was first introduced in 1965 by H. A. Lardy, and initially referred to the liquid that was produced by breaking cells apart and pelleting all the insoluble components by ultracentrifugation. Such a soluble cell extract is not identical to the soluble part of the cell cytoplasm and is usually called a cytoplasmic fraction.

The term *cytosol* is now used to refer to the liquid phase of the cytoplasm in an intact cell. This excludes any part of the cytoplasm that is contained within organelles. Due to the possibility of confusion between the use of the word "cytosol" to refer to both extracts of cells and the soluble part of the cytoplasm in intact cells, the phrase "aqueous cytoplasm" has been used to describe the liquid contents of the cytoplasm of living cells.

Prior to this, other terms were used for the cell fluid, not always synonymously, as its nature was not very clear.

Properties and Composition

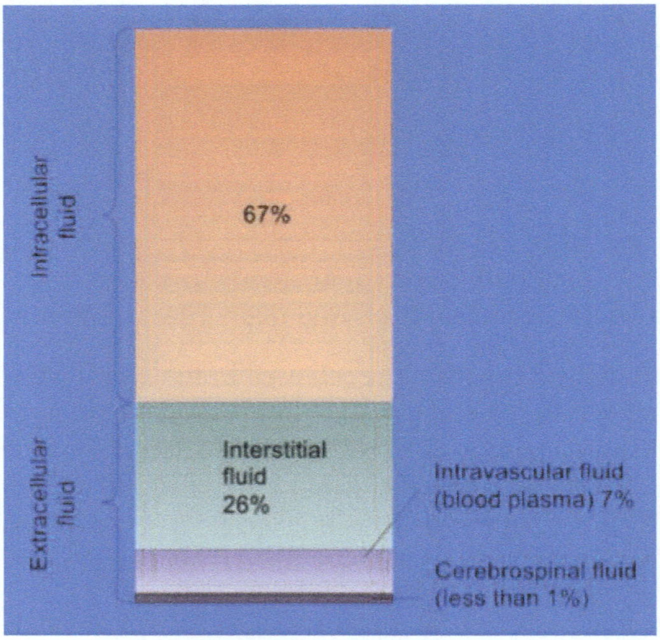

Intracellular fluid content in humans

The proportion of cell volume that is cytosol varies: for example while this compartment forms the bulk of cell structure in bacteria, in plant cells the main compartment is the large central vacuole. The cytosol consists mostly of water, dissolved ions, small molecules, and large water-soluble molecules (such as proteins). The majority of these non-protein molecules have a molecular mass of less than 300 Da. This mixture of small molecules is extraordinarily complex, as the variety of molecules that are involved in metabolism (the metabolites) is immense. For example, up to 200,000 different small molecules might be made in plants, although not all these will be present in the same species, or in a single cell. Estimates of the number of metabolites in single cells such as *E. coli* and baker's yeast predict that under 1,000 are made.

Water

Most of the cytosol is water, which makes up about 70% of the total volume of a typical cell. The pH of the intracellular fluid is 7.4. while human cytosolic pH ranges between 7.0 - 7.4, and is usually higher if a cell is growing. The viscosity of cytoplasm is roughly the same as pure water, although diffusion of small molecules through this liquid is about fourfold slower than in pure water, due mostly to collisions with the large numbers of macromolecules in the cytosol. Studies in the brine shrimp have examined how water affects cell functions; these saw that a 20% reduction in the amount of water in a cell inhibits metabolism, with metabolism decreasing progressively as the cell dries out and all metabolic activity halting when the water level reaches 70% below normal.

Although water is vital for life, the structure of this water in the cytosol is not well understood, mostly because methods such as nuclear magnetic resonance spectroscopy only give information on the average structure of water, and cannot measure local variations at the microscopic scale. Even the structure of pure water is poorly understood, due to the ability of water to form structures such as water clusters through hydrogen bonds.

The classic view of water in cells is that about 5% of this water is strongly bound in by solutes or macromolecules as water of solvation, while the majority has the same structure as pure water. This water of solvation is not active in osmosis and may have different solvent properties, so that some dissolved molecules are excluded, while others become concentrated. However, others argue that the effects of the high concentrations of macromolecules in cells extend throughout the cytosol and that water in cells behaves very differently from the water in dilute solutions. These ideas include the proposal that cells contain zones of low and high-density water, which could have widespread effects on the structures and functions of the other parts of the cell. However, the use of advanced nuclear magnetic resonance methods to directly measure the mobility of water in living cells contradicts this idea, as it suggests that 85% of cell water acts like that pure water, while the remainder is less mobile and probably bound to macromolecules.

Ions

The concentrations of the other ions in cytosol are quite different from those in extracellular fluid and the cytosol also contains much higher amounts of charged macromolecules such as proteins and nucleic acids than the outside of the cell structure.

Typical ion concentrations in mammalian cytosol and blood.

Ion	Concentration in cytosol (millimolar)	Concentration in blood (millimolar)
Potassium	139	4
Sodium	12	145
Chloride	4	116
Bicarbonate	12	29
Amino acids in proteins	138	9
Magnesium	0.8	1.5
Calcium	<0.0002	1.8

In contrast to extracellular fluid, cytosol has a high concentration of potassium ions and a low concentration of sodium ions. This difference in ion concentrations is critical for osmoregulation, since if the ion levels were the same inside a cell as outside, water would enter constantly by osmosis - since the levels of macromolecules inside cells are higher than their levels outside. Instead, sodium ions are expelled and potassium ions taken up by the Na^+/K^+-ATPase, potassium ions then flow down their concentration gradient through potassium-selection ion channels, this loss of positive charge creates a negative membrane potential. To balance this potential difference, negative chloride ions also exit the cell, through selective chloride channels. The loss of sodium and chloride ions compensates for the osmotic effect of the higher concentration of organic molecules inside the cell.

Cells can deal with even larger osmotic changes by accumulating osmoprotectants such as betaines or trehalose in their cytosol. Some of these molecules can allow cells to survive being completely dried out and allow an organism to enter a state of suspended animation called cryptobiosis. In this state the cytosol and osmoprotectants become a glass-like solid that helps stabilize proteins and cell membranes from the damaging effects of desiccation.

The low concentration of calcium in the cytosol allows calcium ions to function as a second messenger in calcium signaling. Here, a signal such as a hormone or an action potential opens calcium channels so that calcium floods into the cytosol. This sudden increase in cytosolic calcium activates other signalling molecules, such as calmodulin and protein kinase C. Other ions such as chloride and potassium may also have signaling functions in the cytosol, but these are not well understood.

Macromolecules

Protein molecules that do not bind to cell membranes or the cytoskeleton are dissolved in the cytosol. The amount of protein in cells is extremely high, and approaches 200 mg/ml, occupying about 20-30% of the volume of the cytosol. However, measuring precisely how much protein is dissolved in cytosol in intact cells is difficult, since some proteins appear to be weakly associated with membranes or organelles in whole cells and are released into solution upon cell lysis. Indeed, in experiments where the plasma membrane of cells were carefully disrupted using saponin, without damaging the other cell membranes, only about one quarter of cell protein was released. These cells were also able to synthesize proteins if given ATP and amino acids, implying that many of the enzymes in cytosol are bound to the cytoskeleton. However, the idea that the majority of the proteins in cells are tightly bound in a network called the microtrabecular lattice is now seen as unlikely.

In prokaryotes the cytosol contains the cell's genome, within a structure known as a nucleoid. This is an irregular mass of DNA and associated proteins that control the transcription and replication of the bacterial chromosome and plasmids. In eukaryotes the genome is held within the cell nucleus, which is separated from the cytosol by nuclear pores that block the free diffusion of any molecule larger than about 10 nanometres in diameter.

This high concentration of macromolecules in cytosol causes an effect called macromolecular crowding, which is when the effective concentration of other macromolecules is increased, since they have less volume to move in. This crowding effect can produce large changes in both the rates and the position of chemical equilibrium of reactions in the cytosol. It is particularly important in its ability to alter dissociation constants by favoring the association of macromolecules, such as when multiple proteins come together to form protein complexes, or when DNA-binding proteins bind to their targets in the genome.

Organization

Although the components of the cytosol are not separated into regions by cell membranes, these components do not always mix randomly and several levels of organization can localize specific molecules to defined sites within the cytosol.

Concentration Gradients

Although small molecules diffuse rapidly in the cytosol, concentration gradients can still be produced within this compartment. A well-studied example of these are the "calcium sparks" that are produced for a short period in the region around an open calcium channel. These are about 2 micrometres in diameter and last for only a few milliseconds, although several sparks can merge to form larger gradients, called "calcium waves". Concentration gradients of other small molecules, such as oxygen and adenosine triphosphate may be produced in cells around clusters of mitochondria, although these are less well understood.

Protein Complexes

Proteins can associate to form protein complexes, these often contain a set of proteins with similar functions, such as enzymes that carry out several steps in the same metabolic pathway. This organization can allow substrate channeling, which is when the product of one enzyme is passed directly to the next enzyme in a pathway without being released into solution. Channeling can make a pathway more rapid and efficient than it would be if the enzymes were randomly distributed in the cytosol, and can also prevent the release of unstable reaction intermediates. Although a wide variety of metabolic pathways involve enzymes that are tightly bound to each other, others may involve more loosely associated complexes that are very difficult to study outside the cell. Consequently, the importance of these complexes for metabolism in general remains unclear.

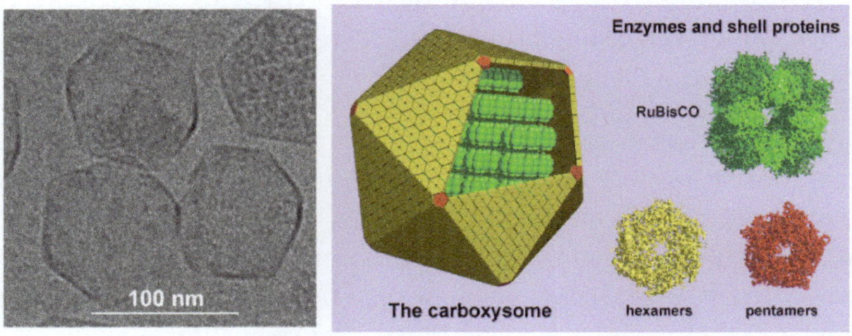

Carboxysomes are protein-enclosed bacterial microcompartments within the cytosol. On the left is an electron microscope image of carboxysomes, and on the right a model of their structure.

Protein Compartments

Some protein complexes contain a large central cavity that is isolated from the remainder of the cytosol. One example of such an enclosed compartment is the proteasome. Here, a set of subunits form a hollow barrel containing proteases that degrade cytosolic proteins. Since these would be damaging if they mixed freely with the remainder of the cytosol, the barrel is capped by a set of regulatory proteins that recognize proteins with a signal directing them for degradation (a ubiquitin tag) and feed them into the proteolytic cavity.

Another large class of protein compartments are bacterial microcompartments, which are made of a protein shell that encapsulates various enzymes. These compartments are typically about 100-200 nanometres across and made of interlocking proteins. A well-understood example is the carboxysome, which contains enzymes involved in carbon fixation such as RuBisCO.

Cytoskeletal Sieving

Although the cytoskeleton is not part of the cytosol, the presence of this network of filaments restricts the diffusion of large particles in the cell. For example, in several studies tracer particles larger than about 25 nanometres (about the size of a ribosome) were excluded from parts of the cytosol around the edges of the cell and next to the nucleus. These "excluding compartments" may contain a much denser meshwork of actin fibres than the remainder of the cytosol. These microdomains could influence the distribution of large structures such as ribosomes and organelles within the cytosol by excluding them from some areas and concentrating them in others.

Function

The cytosol has no single function and is instead the site of multiple cell processes. Examples of these processes include signal transduction from the cell membrane to sites within the cell, such as the cell nucleus, or organelles. This compartment is also the site of many of the processes of cytokinesis, after the breakdown of the nuclear membrane in mitosis. Another major function of cytosol is to transport metabolites from their site of production to where they are used. This is relatively simple for water-soluble molecules, such as amino acids, which can diffuse rapidly through the cytosol. However, hydrophobic molecules, such as fatty acids or sterols, can be transported through the cytosol by specific binding proteins, which shuttle these molecules between cell membranes. Molecules taken into the cell by endocytosis or on their way to be secreted can also be transported through the cytosol inside vesicles, which are small spheres of lipids that are moved along the cytoskeleton by motor proteins.

The cytosol is the site of most metabolism in prokaryotes, and a large proportion of the metabolism of eukaryotes. For instance, in mammals about half of the proteins in the cell are localized to the cytosol. The most complete data are available in yeast, where metabolic reconstructions indicate that the majority of both metabolic processes and metabolites occur in the cytosol. Major metabolic pathways that occur in the cytosol in animals are protein biosynthesis, the pentose phosphate pathway, glycolysis and gluconeogenesis. The localization of pathways can be different in other organisms, for instance fatty acid synthesis occurs in chloroplasts in plants and in apicoplasts in apicomplexa.

References

- Mullins, Christopher (2004). "Theory of Organelle Biogenesis: A Historical Perspective". The Biogenesis of Cellular Organelles. Springer Science+Business Media, National Institutes of Health. ISBN 0-306-47990-7.

- Cremer, Thomas (1985). Von der Zellenlehre zur Chromosomentheorie. Berlin, Heidelberg, New York, Tokyo: Springer Verlag. ISBN 3-540-13987-7. Online Version here
- Lodish, H; Berk A; Matsudaira P; Kaiser CA; Krieger M; Scott MP; Zipursky SL; Darnell J. (2004). Molecular Cell Biology (5th ed.). New York: WH Freeman. ISBN 0-7167-2672-6.
- Lehninger, Albert L.; Nelson, David L.; Cox, Michael M. (2000). Lehninger principles of biochemistry (3rd ed.). New York: Worth Publishers. ISBN 1-57259-931-6.
- Nierhaus, Knud H.; Daniel N. Wilson (2004). Protein Synthesis and Ribosome Structure: Translating the Genome. Wiley-VCH. ISBN 3-527-30638-2.
- Rutledge, Ryan D.; Wright, David W. (2013). "Biomineralization: Peptide-Mediated Synthesis of Materials". In Lukehart, C.M.; Scott, R.A. Nanomaterials: Inorganic and Bioinorganic Perspectives. EIC Books. Wiley. ISBN 978-1-118-62522-4. Retrieved 2016-03-14.
- Campbell, Neil A.; Reece, Jane B.; Urry, Lisa A.; Cain, Michael L.; Wasserman, Steven A.; Minorsky, Peter V.; Jackson, Robert B. (2008). Biology (8th ed.). p. 119. ISBN 978-0-8053-6844-4.
- Hudler, George W. (1998). Magical Mushrooms, Mischievous Molds. Princeton, NJ: Princeton University Press, 7. ISBN 0-691-02873-7.
- Alexopoulos, C. J., C. W. Mims, & M. Blackwell (1996). Introductory Mycology 4. New York: John Wiley & Sons, 687-688. ISBN 0-471-52229-5.
- Brock, Thomas D., Michael T. Madigan, John M. Martinko, & Jack Parker. (1994) Biology of Microorganisms, 7th ed., pages 818-819, 824 (Englewood Cliffs, NJ: Prentice Hall). ISBN 0-13-042169-3.
- Hardin, Jeff; Bertoni, Gregory; Kleinsmith, Lewis J. (2015). Becker's World of the Cell (8th ed.). New York: Pearson. pp. 422–446. ISBN 978013399939-6.
- McKinley, Michael; Dean O'Loughlin, Valerie; Pennefather-O'Brien, Elizabeth; Harris, Ronald (2015). Human Anatomy (4th Ed.). New York: McGraw Hill Education. p. 29. ISBN 0-07-352573-1.
- Paulin D, Li Z (Nov 2004). "Desmin: a major intermediate filament protein essential for the structural integrity and function of muscle". Experimental Cell Research. 301 (1): 1–7. doi:10.1016/j.yexcr.2004.08.004. PMID 15501438. Retrieved 17 April 2015.
- Gunning PW, Ghoshdastider U, Whitaker S, Popp D, Robinson RC (2015). "The evolution of compositionally and functionally distinct actin filaments". Journal of Cell Science 128 (11): 2009–19. doi:10.1242/jcs.165563. PMID 25788699.

4

Cell Biology: Theories and Processes

Our understanding of cells and cell function come from theories that were promulgated by scholars through the years. In order to study cell and cell structure in detail, it is vital to know the basic stages of the cell cycle as well as theories on its origin. This chapter strategically encompasses and incorporates the major components and key concepts of cell biology, providing a complete understanding.

Cell Disruption

Cell disruption is a method or process for releasing biological molecules from inside a cell.

Laboratory cell disruptor

Methods

The production of biologically interesting molecules using cloning and culturing methods allows the study and manufacture of relevant molecules. Except for excreted molecules, cells producing molecules of interest must be disrupted. This page discusses various methods.

Bead Method

A common laboratory-scale mechanical method for cell disruption uses tiny glass, ceramic or steel beads mixed with a sample suspended in aqueous media. First developed by Tim Hopkins in the late 1970s, the sample and bead mix is subjected to high level agitation by stirring or shaking. Beads collide with the cellular sample, cracking open the cell to release intercellular components. Unlike some other methods, mechanical shear is moderate during homogenization resulting in excellent membrane or subcellular preparations. The method, often called "beadbeating", works well for all types of cellular material - from spores to animal and plant tissues. It is the most widely used method of yeast lysis, and can yield breakage of over 50%. It has the advantage over other mechanical cell disruption methods of being able to disrupt very small sample sizes, process many samples at a time with no cross-contamination concerns, and does not release potentially harmful aerosols in the process.

In the simplest example of the method, an equal volume of beads are added to a cell or tissue suspension in a test tube and the sample is vigorously mixed on a common laboratory vortex mixer. While processing times are slow, taking 3-10 times longer than that in specialty shaking machines, it works well for easily disrupted cells and is inexpensive.

In most laboratories, beadbeating is done in sealed, plastic vials, centrifuge tubes, or deep-well microtiter plates. The sample and tiny beads are agitated at about 2000 oscillations per minute in specially designed vial shakers driven by high power electric motors. Cell disruption is complete in 1–3 minutes of shaking. Machines are available that can process hundreds of samples simultaneously inside deep well microplates.

Successful beadbeating is dependent not only design features of the shaking machine (which take into consideration shaking oscillations per minute, shaking throw or distance, shaking orientation and vial orientation), but also the selection of correct bead size (0.1–6 mm diameter), bead composition (glass, ceramic, steel) and bead load in the vial.

All high energy beadbeating machines warm the sample about 10 degrees/minute. This is due to frictional collisions of the beads during homogenization. Cooling of the sample during or after beadbeating may be necessary to prevent damage to heat sensitive proteins such as enzymes. Sample warming can be controlled by beadbeating for short time intervals with cooling on ice between each interval, by processing vials in pre-chilled aluminum vial holders or by circulating gaseous coolant through the machine during beadbeating.

A different beadbeater configuration, suitable for larger sample volumes, uses a fluorocarbon rotor inside a 15, 50 or 200 ml chamber to agitate the beads. In this configuration, the chamber can be surrounded by a static cooling jacket. Using the same rotor/chamber configuration, large commercial machines are available to process many liters of cell suspension. Currently, these machines are limited to processing monocellular organisms such as yeast, algae and bacteria.

Cryopulverization

Samples with a tough extracellular matrix, such as animal connective tissue, some tumor biopsy samples, venous tissue, cartilage, seeds, etc., are reduced to a fine powder by impact pulverization at liquid nitrogen temperatures. This technique, known as cryopulverization, is based on the fact that biological samples containing a significant fraction of water become brittle at extremely cold temperatures. This technique was first described by Smucker and Pfister in 1975, who referred to the technique as cryo-impacting. The authors demonstrated cells are effectively broken by this method, confirming by phase and electron microscopy that breakage planes cross cell walls and cytoplasmic membranes.

The technique can done using a mortar and pestle cooled to liquid nitrogen temperatures, but use of this classic apparatus is laborious and sample loss is often a concern. Specialised stainless steel pulverizers generically known as Tissue Pulverizers are also available for this purpose. They require less manual effort, give good sample recovery and are easy to clean between samples. Advantages of this technique are higher yields of proteins and nucleic acids from small, hard tissue samples - especially when used as a preliminary step to mechanical or chemical/solvent cell disruption methods mentioned above.

Microfluidizer Method

This method used for cell disruption strongly influences the physicochemical properties of the lysed cell suspension, such as particle size, viscosity, protein yield and enzyme activity. In recent years the Microfluidizer method has gained popularity in cell disruption due to its ease of use and efficiency at disrupting many different kinds of cells. The Microfluidizer technology was licensed from a company called Arthur D. Little and was first developed and utilized in the 1980s, initially starting as a tool for liposome creation. It has since been used in other applications such as cell disruption nanoemulsions, and solid particle size reduction, among others.

By using microchannels with fixed geometry, and an intensifier pump, high shear rates are generated that rupture the cells. This method of cell lysis can yield breakage of over 90% of E. coli cells.

Many proteins are extremely temperature-sensitive, and in many cases can start to denature at temperatures of only 4 degrees celsius. Within the microchannels, temperatures exceed 4 degrees celsius, but the machine is designed to cool quickly so that the time the cells are exposed to elevated temperatures is extremely short (residence time 25 ms-40 ms). Because of this effective temperature control, the Microfluidizer yields higher levels of active proteins and enzymes than other mechanical methods when the proteins are temperature-sensitive.

Viscosity changes are also often observed when disrupting cells. If the cell suspension viscosity is high, it can make downstream handling—such as filtration and accurate pipetting—quite difficult. The viscosity changes observed with a Microfluidizer are relatively low, and decreases with further additional passes through the machine.

In contrast to other mechanical disruption methods the Microfluidizer breaks the cell membranes efficiently but gently, resulting in relatively large cell wall fragments (450 nm), and thus making it easier to separate the cell contents. This can lead to shorter filtration times and better centrifugation separation.

Microfluidizer technology scales from one milliliter to thousands of liters.

Nitrogen Decompression

For nitrogen decompression, large quantities of nitrogen are first dissolved in the cell under high pressure within a suitable pressure vessel. Then, when the gas pressure is suddenly released, the nitrogen comes out of the solution as expanding bubbles that stretch the membranes of each cell until they rupture and release the contents of the cell.

Nitrogen decompression is more protective of enzymes and organelles than ultrasonic and mechanical homogenizing methods and compares favorably to the controlled disruptive action obtained in a PTFE and glass mortar and pestle homogenizer. While other disruptive methods depend upon friction or a mechanical shearing action that generate heat, the nitrogen decompression procedure is accompanied by an adiabatic expansion that cools the sample instead of heating it.

The blanket of inert nitrogen gas that saturates the cell suspension and the homogenate offers protection against oxidation of cell components. Although other gases: carbon dioxide, nitrous oxide, carbon monoxide and compressed air have been used in this technique, nitrogen is preferred because of its non-reactive nature and because it does not alter the pH of the suspending medium. In addition, nitrogen is preferred because it is generally available at low cost and at pressures suitable for this procedure.

Once released, subcellular substances are not exposed to continued attrition that might denature the sample or produce unwanted damage. There is no need to watch for a peak between enzyme activity and percent disruption. Since nitrogen bubbles are generated within each cell, the same disruptive force is applied uniformly throughout the sample, thus ensuring unusual uniformity in the product. Cell-free homogenates can be produced.

The technique is used to homogenize cells and tissues, release intact organelles, prepare cell membranes, release labile biochemicals, and produce uniform and repeatable homogenates without subjecting the sample to extreme chemical or physical stress.

The method is particularly well suited for treating mammalian and other membrane-bound cells. It has also been used successfully for treating plant cells, for releasing virus from fertilized eggs and for treating fragile bacteria. It is not recommended for untreated bacterial cells. Yeast, fungus, spores and other materials with tough cell walls do not respond well to this method.

Cell Theory

In biology, cell theory is a scientific theory which describes the properties of cells. These cells are the basic unit of structure in all organisms and also the basic unit of reproduction. With continual improvements made to microscopes over time, magnification technology advanced enough to discover cells in the 17th century. This discovery is largely attributed to Robert Hooke, and began the scientific study of cells, also known as cell biology. Over a century later, many debates about cells began amongst scientists. Most of these debates involved the nature of cellular regeneration, and the idea of cells as a fundamental unit of life. Cell theory was eventually formulated in 1838. This is usually credited to Matthias Schleiden and Theodor Schwann. However, many other scientists like Rudolf Virchow contributed to the theory. Cell theory has become the foundation of biology and is the most widely accepted explanation of the function of cells.

The three tenets to the cell theory are as described below:

1. All living organisms are composed of one or more cells. (However, this is controversial because non-cellular life such as viruses are disputed as a life form).

2. The cell is the basic unit of structure and organization in organisms.
3. Cells come from pre-existing cells.

Microscopes

Anton van Leeuwenhoek's microscope from the 17th century with a magnification of 270x.

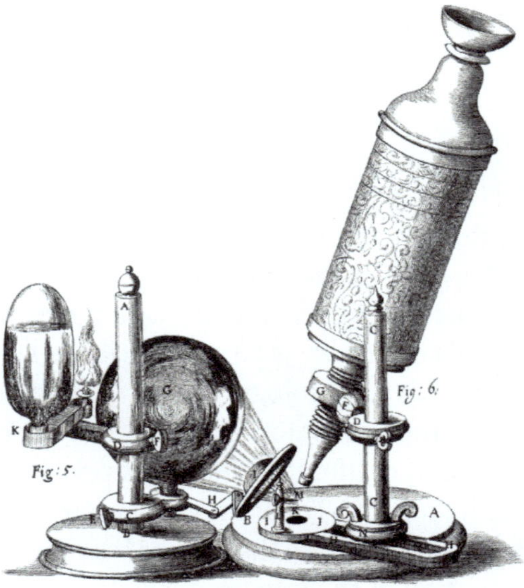

Robert Hooke's microscope

The discovery of the cell was made possible through the invention of the microscope. In the first century BC, Romans were able to make glass, discovering that objects appeared to be larger under the glass. In Italy during the 12th century, Salvino D'Armate made a piece of glass fit over one eye, allowing for a magnification effect to that eye. It was not until the 1590s when a Dutch spectacle maker Zacharias Jansen began to test lenses that progress had been made to microscopes. Jansen was able to obtain about 9x magnification, but the objects appeared to be blurry. In 1595, Jansen and his father built the first compound microscope. While simple glasses were able to magnify objects, they were not considered to be a microscope. A compound microscope was defined by having two or more lenses in a hollow tube. In 1665, Robert Hooke used a microscope about six inches long with two convex lenses inside and examined specimens under reflected light for the observations in his book *Micrographia*. Hooke also used a simpler microscope with a single lens for examining specimens with directly transmitted light, because this allowed for a clearer image.

However, the first real invention and use of the microscope were done by Anton van Leeuwenhoek. He was a Dutch draper that took the interest in microscopes after seeing one while on an apprenticeship in Amsterdam in 1648. At some point in his life before 1668, he was able to learn how to grind lenses. This eventually led to Leeuwenhoek making his own unique microscope. His were instead simple powerful magnifying glasses, rather than a compound microscope. This was because he was able to use a single lens that was a small glass sphere but allowed for a magnification of 270x. This was a large progression since the magnification before was only a maximum of 50x. After Leeuwenhoek, there was not much progress for the microscopes until the 1850s, two hundred years later. Carl Zeiss, a German engineer who manufactured microscopes, began to make changes to the lenses used. But the optical quality did not improve until the 1880s when he hired Otto Schott and eventually Ernst Abbe.

These microscopes could focus on objects the size of a wavelength or larger, giving restrictions still to advancement in discoveries with objects smaller than a wavelength. Later in the 1920s, the electron microscope was developed, making it possible to view objects that are smaller than a wavelength, once again, changing the possibilities in science.

Discovery of Cells

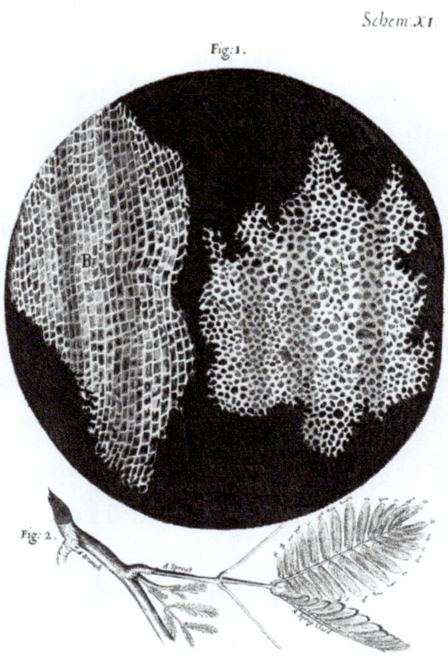

Drawing of the structure of cork by Robert Hooke that appeared in *Micrographia*.

The cell was first discovered by Robert Hooke in 1665, which can be found to be described in his book Micrographia. In this book, he gave 60 'observations' in detail of various objects under a coarse, compound microscope. One observation was from very thin slices of bottle cork. Hooke discovered a multitude of tiny pores that he named "cells". This came from the Latin word Cella, meaning 'a small room' like monks lived in and also Cellulae, which meant the six sided cell of a honeycomb. However, Hooke did not know their real structure or function. What Hooke had thought were cells, were actually empty cell walls of plant tissues. With microscopes during this time having a low magnification, Hooke was unable to see that there were other internal components to the cells he was observing. Therefore, he did not think the "cellulae" were alive. His cell observations gave no indication of the nucleus and other organelles found in most living cells. In Micrographia, Hooke also observed mould, bluish in color, found on leather. After studying it under his microscope, he was unable to observe "seeds" that would have indicated how the mould was multiplying in quantity. This led to Hooke suggesting that spontaneous generation, from either natural or artificial heat, was the cause. Since this was an old Aristotelian theory still accepted at the time, others did not reject it and was not disproved until Leeuwenhoek later discovers generation is achieved otherwise.

Anton van Leeuwenhoek is another scientist who saw these cells soon after Hooke did. He made use of a microscope containing improved lenses that could magnify objects almost 300-fold, or 270x. Under these microscopes, Leeuwenhoek found motile objects. In a letter to The Royal Society on October 9, 1676, he states that motility is a quality of life therefore these were living organisms. Over time, he wrote many more papers in which described many specific forms of microorganisms. Leeuwenhoek named these "animalcules," which included protozoa and other unicellular organisms, like bacteria. Though he did not have much formal education, he was able to identify the first accurate description of red blood cells and discovered bacteria after gaining interest in the sense of taste that resulted in Leeuwenhoek to observe the tongue of an ox, then leading him to study "pepper water" in 1676. He also found for the first time the sperm cells of animals and humans. Once discovering these types of cells, Leeuwenhoek saw that the fertilization process requires the sperm cell to enter the egg cell. This put an end to the previous theory of spontaneous generation. After reading letters by Leeuwenhoek, Hooke was the first to confirm his observations that were thought to be unlikely by other contemporaries.

The cells in animal tissues were observed after plants were because the tissues were so fragile and susceptible to tearing, it was difficult for such thin slices to be prepared for studying. Biologists believed that there was a fundamental unit to life, but were unsure what this was. It would not be until over a hundred years later that this fundamental unit was connected to cellular structure and existence of cells in animals or plants. This conclusion was not made until Henri Dutrochet. Besides stating "the cell is the fundamental element of organization", Dutrochet also claimed that cells were not just a structural unit, but also a physiological unit.

In 1804, Karl Rudolphi and J.H.F. Link were awarded the prize for "solving the problem of the nature of cells", meaning they were the first to prove that cells had independent cell walls by the Königliche Societät der Wissenschaft (Royal Society of Science), Göttingen. Before, it had been thought that cells shared walls and the fluid passed between them this way.

Cell Theory

Matthias Jakob Schleiden (1804–1881)

Theodor Schwann (1810–1882)

Credit for developing cell theory is usually given to two scientists: Theodor Schwann and Matthias Jakob Schleiden. While Rudolf Virchow contributed to the theory, he is not as credited for his attributions toward it. In 1838, Schleiden suggested that every structural part of a plant was made up of cells or the result of cells. He also suggested that cells were made by a crystallization process either within other cells or from the outside. However, this was not an original idea of Schlieden. He claimed this theory as his own, though Barthelemy Dumortier had stated it years before him. This crystallization process is no longer accepted with modern cell theory. In 1839, Theodor Schwann states that along with plants, animals are composed of cells or the product of cells in their structures. This was a major advancement in the field of biology since little was known about animal structure up to this point compared to plants. From these conclusions about plants and animals, two of the three tenets of cell theory were postulated.

1. All living organisms are composed of one or more cells

2. The cell is the most basic unit of life

Schleiden's theory of free cell formation through crystallization was refuted in the 1850s by Robert Remak, Rudolf Virchow, and Albert Kolliker. In 1855, Rudolf Virchow added the third tenet to cell theory. In Latin, this tenet states *Omnis cellula e cellula*. This translated to:

3. All cells arise only from pre-existing cells

However, the idea that all cells come from pre-existing cells had in fact already been proposed by Robert Remak; it has been suggested that Virchow plagiarized Remak and did not give him credit. Remak published observations in 1852 on cell division, claiming Schleiden and Schawnn were incorrect about generation schemes. He instead said that binary fission, which was first introduced by Dumortier, was how reproduction of new animal cells were made. Once this tenet was added, the classical cell theory was complete.

Modern Interpretation

The generally accepted parts of modern cell theory include:

1. All known living things are made up of one or more cells
2. All living cells arise from pre-existing cells by division.
3. The cell is the fundamental unit of structure and function in all living organisms.
4. The activity of an organism depends on the total activity of independent cells.
5. Energy flow (metabolism and biochemistry) occurs within cells.
6. Cells contain DNA which is found specifically in the chromosome and RNA found in the cell nucleus and cytoplasm.
7. All cells are basically the same in chemical composition in organisms of similar species .

The Modern Version of the Cell Theory

The modern version of the cell theory includes the ideas that:

- Energy flow occurs within cells.
- Heredity information (DNA) is passed on from cell to cell.
- All cells have the same basic chemical composition.

Opposing Concepts in Cell Theory: History And Background

The cell was first discovered by Robert Hooke in 1665 using a microscope. The first cell theory is credited to the work of Theodor Schwann and Matthias Jakob Schleiden in the 1830s. In this theory the internal contents of cells were called protoplasm and described as a jelly-like substance, sometimes called living jelly. At about the same time, colloidal chemistry began its development, and the concepts of bound water emerged. A colloid being something between a solution and a suspension, where Brownian motion is sufficient to prevent sedimentation. The idea of a semi-permeable membrane, a barrier that is permeable to solvent but impermeable to solute molecules was developed at about the same time. The term osmosis originated in 1827 and its importance to physiological phenomena realized, but it wasn't until 1877, when the botanist Pfeffer proposed the membrane theory of cell physiology. In this view, the cell was seen to be enclosed by a thin surface, the plasma membrane, and cell water and solutes such as a potassium ion existed in a physical state like that of a dilute solution. In 1889 Hamburger used hemolysis of erythrocytes to determine the permeability of various solutes. By measuring the time required for the cells to swell past their elastic limit, the rate at which solutes entered the cells could be estimated by the accompanying change in cell volume. He also found that there was an apparent nonsolvent volume of about 50% in red blood cells and later showed that this includes water of hydration in addition to the protein and other nonsolvent components of the cells.

Evolution of the Membrane and Bulk Phase Theories

Two opposing concepts developed within the context of studies on osmosis, permeability, and electrical properties of cells. The first held that these properties all belonged to the plasma membrane whereas the other predominant view was that the protoplasm was responsible for these properties. The membrane theory developed as a succession of ad-hoc additions and changes to the theory to overcome experimental hurdles. Overton (a distant cousin of Charles Darwin) first proposed the concept of a lipid (oil) plasma membrane in 1899. The major weakness of the lipid membrane was the lack of an explanation of the high permeability to water, so Nathansohn (1904) proposed the mosaic theory. In this view, the membrane is not a pure lipid layer, but a mosaic of areas with lipid and areas with semipermeable gel. Ruhland refined the mosaic theory to include pores to allow additional passage of small molecules. Since membranes are generally less permeable to anions, Leonor Michaelis concluded that ions are adsorbed to the walls of the pores, changing the permeability of the pores to ions by electrostatic repulsion. Michaelis demonstrated the membrane potential (1926) and proposed that it was related to the distribution of ions across the membrane. Harvey and Danielli (1939) proposed a lipid bilayer membrane covered on each side with a layer of protein to account for measurements of surface tension. In 1941 Boyle & Conway showed that the membrane of frog muscle was permeable to both K+ and Cl-, but apparently not to Na+, so the idea of electrical charges in the pores was unnecessary since a single critical pore size would explain the permeability to K+, H+, and Cl- as well as the impermeability to Na+, Ca+, and Mg++. Over the same time period, it was shown (Procter & Wilson, 1916) that gels, which do not have a semipermeable membrane, would swell in dilute solutions. Loeb (1920) also studied gelatin extensively, with and without a membrane, showing that more of the properties attributed to the plasma membrane could be duplicated in gels without a membrane. In particular, he found that an electrical potential difference between the gelatin and the outside medium could be developed, based on the H+ concentration. Some criticisms of the membrane theory developed in the 1930s, based on observations such as the ability of some cells to swell and increase their surface area by a factor of 1000. A lipid layer cannot stretch to that extent without becoming a patchwork (thereby losing its barrier properties. Such criticisms stimulated continued studies on protoplasm as the principal agent determining cell permeability properties. In 1938, Fischer and Suer proposed that water in the protoplasm is not free but in a chemically combined form—the protoplasm represents a combination of protein, salt and water—and demonstrated the basic similarity between swelling in living tissues and the swelling of gelatin and fibrin gels. Dimitri Nasonov (1944) viewed proteins as the central components responsible for many properties of the cell, including electrical properties. By the 1940s, the bulk phase theories were not as well developed as the membrane theories. In 1941, Brooks & Brooks published a monograph The Permeability of Living Cells, which rejects the bulk phase theories.

The Emergence of the steady-state Membrane Pump Concept

With the development of radioactive tracers, it was shown that cells are not impermeable to Na^+. This was difficult to explain with the membrane barrier theory, so the sodium pump was proposed to continually remove Na+ as it permeates cells. This drove the concept that cells are in a state of dynamic equilibrium, constantly using energy to maintain ion gradients. In 1935, Karl Lohmann discovered ATP and its role as a source of energy for cells, so the concept of a metabolically-driven sodium pump was proposed. The tremendous success of Hodgkin, Huxley, and Katz in the development of the membrane theory of cellular membrane potentials, with differential equations that modeled the phenomena correctly, provided even more support for the membrane pump hypothesis.

The modern view of the plasma membrane is of a fluid lipid bilayer that has protein components embedded within it. The structure of the membrane is now known in great detail, including 3D models of many of the hundreds of different proteins that are bound to the membrane. These major developments in cell physiology placed the membrane theory in a position of dominance and stimulated the imagination of most physiologists, who now apparently accept the theory as fact—there are, however, a few dissenters.

The Reemergence of the Bulk Phase Theories

In 1956, Afanasy S. Troshin published a book, *The Problems of Cell Permeability*, in Russian (1958 in German, 1961 in Chinese, 1966 in English) in which he found that permeability was of secondary importance in determination of the patterns of equilibrium between the cell and its environment. Troshin showed that cell water decreased in solutions of galactose or urea although these compounds did slowly permeate cells. Since the membrane theory requires an impermanent solute to sustain cell shrinkage, these experiments cast doubt on the theory. Others questioned whether the cell has enough energy to sustain the sodium/potassium pump. Such questions became even more urgent as dozens of new metabolic pumps were added as new chemical gradients were discovered.

In 1962, Gilbert Ling became the champion of the bulk phase theories and proposed his association-induction hypothesis of living cells.

Types of Cells

Cells can be subdivided into the following subcategories:

1. *Prokaryotes*: Prokaryotes are relatively small cells surrounded by the plasma membrane, with a characteristic cell wall that may differ in composition depending on the particular organism. Prokaryotes lack a nucleus (although they do have circular or linear DNA) and

other membrane-bound organelles (though they do contain ribosomes). The protoplasm of a prokaryote contains the chromosomal region that appears as fibrous deposits under the microscope, and the cytoplasm. Bacteria and Archaea are the two domains of prokaryotes.

2. *Eukaryotes*: Eukaryotic cells are also surrounded by the plasma membrane, but on the other hand, they have distinct nuclei bound by a nuclear membrane or envelope. Eukaryotic cells also contain membrane-bound organelles, such as (mitochondria, chloroplasts, lysosomes, rough and smooth endoplasmic reticulum, vacuoles). In addition, they possess organized chromosomes which store genetic material.

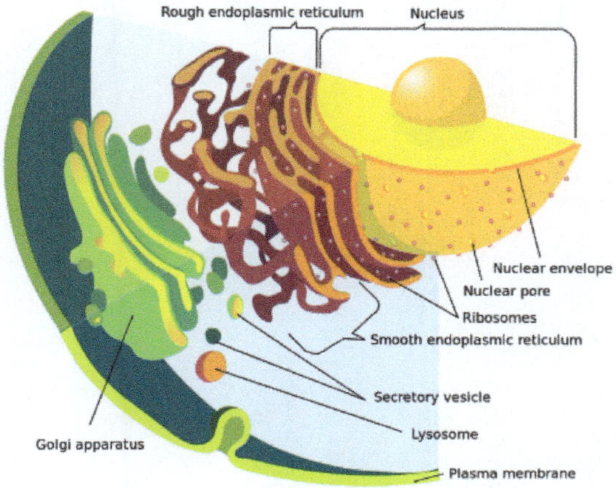

Eukaryote cell.

Animals have evolved a greater diversity of cell types in a multicellular body (100–150 different cell types), compared with 10–20 in plants, fungi, and protoctists.

Cell Culture

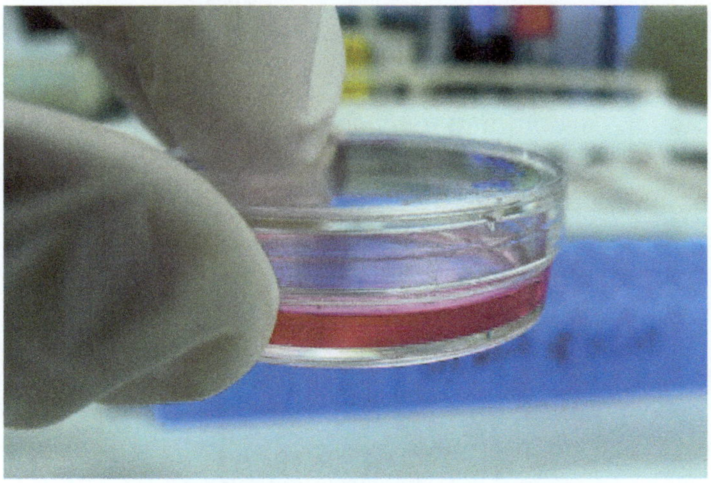

Cell culture in a special tissue culture dish

Cell culture is the process by which cells are grown under controlled conditions, generally outside of their natural environment. In practice, the term "cell culture" now refers to the culturing of cells derived from multicellular eukaryotes, especially animal cells, in contrast with other types of culture that also grow cells, such as plant tissue culture, fungal culture, and microbiological culture (of microbes). The historical development and methods of cell culture are closely interrelated to those of tissue culture and organ culture. Viral culture is also related, with cells as hosts for the viruses.

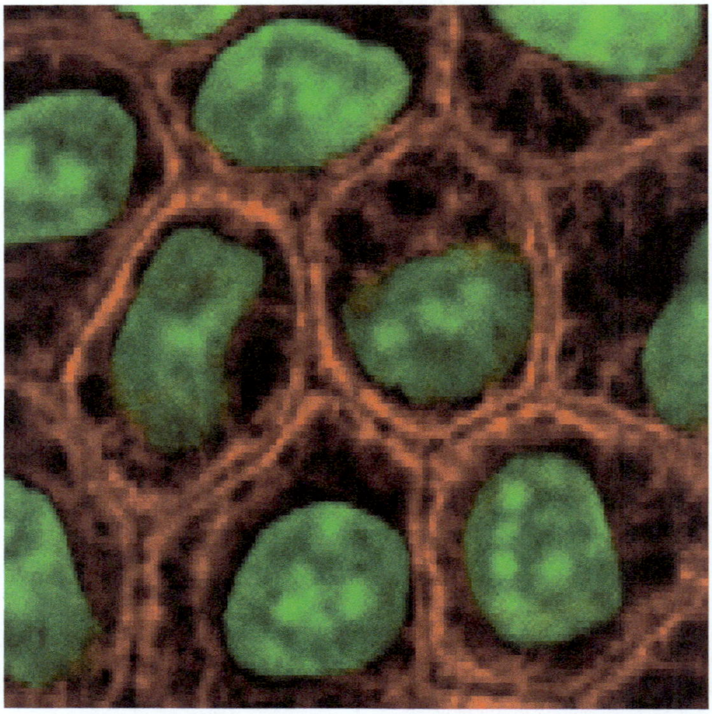

Epithelial cells in culture, stained for keratin (red) and DNA (green)

The laboratory technique of maintaining live cell lines (a population of cells descended from a single cell and containing the same genetic makeup) separated from their original tissue source became more robust in the middle 20th century.

History

The 19th-century English physiologist Sydney Ringer developed salt solutions containing the chlorides of sodium, potassium, calcium and magnesium suitable for maintaining the beating of an isolated animal heart outside of the body. In 1885, Wilhelm Roux removed a portion of the medullary plate of an embryonic chicken and maintained it in a warm saline solution for several days, establishing the principle of tissue culture. Ross Granville Harrison, working at Johns Hopkins Medical School and then at Yale University, published results of his experiments from 1907 to 1910, establishing the methodology of tissue culture.

Cell culture techniques were advanced significantly in the 1940s and 1950s to support research in virology. Growing viruses in cell cultures allowed preparation of purified viruses for the manufacture of vaccines. The injectable polio vaccine developed by Jonas Salk was one of the first products mass-produced using cell culture techniques. This vaccine was made possible by the cell culture research of John Franklin Enders, Thomas Huckle Weller, and Frederick Chapman Robbins, who were awarded a Nobel Prize for their discovery of a method of growing the virus in monkey kidney cell cultures.

Concepts in Mammalian Cell Culture

Isolation of Cells

Cells can be isolated from tissues for *ex vivo* culture in several ways. Cells can be easily purified from blood; however, only the white cells are capable of growth in culture. Mononuclear cells can be released from soft tissues by enzymatic digestion with enzymes such as collagenase, trypsin, or pronase, which break down the extracellular matrix. Alternatively, pieces of tissue can be placed in growth media, and the cells that grow out are available for culture. This method is known as explant culture.

Cells that are cultured directly from a subject are known as primary cells. With the exception of some derived from tumors, most primary cell cultures have limited lifespan.

An established or immortalized cell line has acquired the ability to proliferate indefinitely either through random mutation or deliberate modification, such as artificial expression of the telomerase gene. Numerous cell lines are well established as representative of particular cell types.

Maintaining Cells in Culture

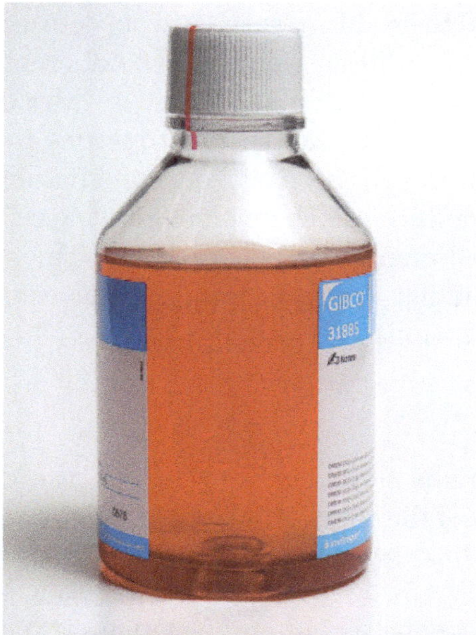

A bottle of DMEM cell culture medium

For the majority of isolated primary cells, they undergo the process of senescence and stop dividing after a certain number of population doublings while generally retaining their viability (described as the Hayflick limit).

Cells are grown and maintained at an appropriate temperature and gas mixture (typically, 37 °C, 5% CO_2 for mammalian cells) in a cell incubator. Culture conditions vary widely for each cell type, and variation of conditions for a particular cell type can result in different phenotypes.

Aside from temperature and gas mixture, the most commonly varied factor in culture systems is the cell growth medium. Recipes for growth media can vary in pH, glucose concentration, growth factors, and the presence of other nutrients. The growth factors used to supplement media are often derived from the serum of animal blood, such as fetal bovine serum (FBS), bovine calf serum, equine serum, and porcine serum. One complication of these blood-derived ingredients is the potential for contamination of the culture with viruses or prions, particularly in medical biotechnology applications. Current practice is to minimize or eliminate the use of these ingredients wherever possible and use human platelet lysate (hPL). This eliminates the worry of cross-species contamination when using FBS with human cells. hPL has emerged as a safe and reliable alternative as a direct replacement for FBS or other animal serum. In addition, chemically defined media can be used to eliminate any serum trace (human or animal), but this cannot always be accomplished with different cell types. Alternative strategies involve sourcing the animal blood from countries with minimum BSE/TSE risk, such as The United States, Australia and New Zealand, and using purified nutrient concentrates derived from serum in place of whole animal serum for cell culture.

Plating density (number of cells per volume of culture medium) plays a critical role for some cell types. For example, a lower plating density makes granulosa cells exhibit estrogen production, while a higher plating density makes them appear as progesterone-producing theca lutein cells.

Cells can be grown either in suspension or adherent cultures. Some cells naturally live in suspension, without being attached to a surface, such as cells that exist in the bloodstream. There are also cell lines that have been modified to be able to survive in suspension cultures so they can be grown to a higher density than adherent conditions would allow. Adherent cells require a surface, such as tissue culture plastic or microcarrier, which may be coated with extracellular matrix (such as collagen and laminin) components to increase adhesion properties and provide other signals needed for growth and differentiation. Most cells derived from solid tissues are adherent. Another type of adherent culture is organotypic culture, which involves growing cells in a three-dimensional (3-D) environment as opposed to two-dimensional culture dishes. This 3D culture system is biochemically and physiologically more similar to *in vivo* tissue, but is technically challenging to maintain because of many factors (e.g. diffusion).

Cell line Cross-Contamination

Cell line cross-contamination can be a problem for scientists working with cultured cells. Studies suggest anywhere from 15–20% of the time, cells used in experiments have been misidentified or contaminated with another cell line. Problems with cell line cross-contamination have even been detected in lines from the NCI-60 panel, which are used routinely for drug-screening studies. Major cell line repositories, including the American Type Culture Collection (ATCC), the European Collection of Cell Cultures (ECACC) and the German Collection of Microorganisms and Cell Cultures (DSMZ), have received cell line submissions from researchers that were misidentified by them. Such contamination poses a problem for the quality of research produced using cell culture lines, and the major repositories are now authenticating all cell line submissions. ATCC uses short tandem repeat (STR) DNA fingerprinting to authenticate its cell lines.

To address this problem of cell line cross-contamination, researchers are encouraged to authenticate their cell lines at an early passage to establish the identity of the cell line. Authentication should be repeated before freezing cell line stocks, every two months during active culturing and before any publication of research data generated using the cell lines. Many methods are used to identify cell lines, including isoenzyme analysis, human lymphocyte antigen (HLA) typing, chromosomal analysis, karyotyping, morphology and STR analysis.

One significant cell-line cross contaminant is the immortal HeLa cell line.

Other Technical Issues

As cells generally continue to divide in culture, they generally grow to fill the available area or volume. This can generate several issues:

- Nutrient depletion in the growth media
- Changes in pH of the growth media
- Accumulation of apoptotic/necrotic (dead) cells
- Cell-to-cell contact can stimulate cell cycle arrest, causing cells to stop dividing, known as contact inhibition.
- Cell-to-cell contact can stimulate cellular differentiation.
- Genetic and epigenetic alterations, with a natural selection of the altered cells potentially leading to overgrowth of abnormal, culture-adapted cells with decreased differentiation and increased proliferative capacity.

Manipulation of Cultured Cells

Among the common manipulations carried out on culture cells are media changes, passaging cells, and transfecting cells. These are generally performed using tissue culture methods that rely on aseptic technique. Aseptic technique aims to avoid contamination with bacteria, yeast, or other cell lines. Manipulations are typically carried out in a biosafety hood or laminar flow cabinet to exclude contaminating micro-organisms. Antibiotics (e.g. penicillin and streptomycin) and antifungals (e.g.amphotericin B) can also be added to the growth media.

As cells undergo metabolic processes, acid is produced and the pH decreases. Often, a pH indicator is added to the medium to measure nutrient depletion.

Media Changes

In the case of adherent cultures, the media can be removed directly by aspiration, and then is replaced. Media changes in non-adherent cultures involve centrifuging the culture and resuspending the cells in fresh media.

Passaging Cells

Passaging (also known as subculture or splitting cells) involves transferring a small number of cells into a new vessel. Cells can be cultured for a longer time if they are split regularly, as it avoids the senescence associated with prolonged high cell density. Suspension cultures are easily passaged with a small amount of culture containing a few cells diluted in a larger volume of fresh media. For adherent cultures, cells first need to be detached; this is commonly done with a mixture of trypsin-EDTA; however, other enzyme mixes are now available for this purpose. A small number of detached cells can then be used to seed a new culture. Some cell cultures, such as RAW cells are mechanically scraped from the surface of their vessel with rubber scrapers.

Transfection and Transduction

Another common method for manipulating cells involves the introduction of foreign DNA by transfection. This is often performed to cause cells to express a gene of interest. More recently, the transfection of RNAi constructs have been realized as a convenient mechanism for suppressing the expression of a particular gene/protein. DNA can also be inserted into cells using viruses, in methods referred to as transduction, infection or transformation. Viruses, as parasitic agents, are well suited to introducing DNA into cells, as this is a part of their normal course of reproduction.

Established Human Cell Lines

Cell lines that originate with humans have been somewhat controversial in bioethics, as they may outlive their parent organism and later be used in the discovery of lucrative medical treatments. In the pioneering decision in this area, the Supreme Court of California held in *Moore v. Regents of the University of California* that human patients have no property rights in cell lines derived from organs removed with their consent.

It is possible to fuse normal cells with an immortalised cell line. This method is used to produce monoclonal antibodies. In brief, lymphocytes isolated from the spleen (or possibly blood) of an immunised animal are combined with an immortal myeloma cell line (B cell lineage) to produce a hybridoma which has the antibody specificity of the primary lymphocyte and the immortality of the myeloma. Selective growth medium (HA or HAT) is used to select against unfused myeloma cells; primary lymphoctyes die quickly in culture and only the fused cells survive. These are screened for production of the required antibody, generally in pools to start with and then after single cloning.

Cell strains

A cell strain is derived either from a primary culture or a cell line by the selection or cloning of cells having specific properties or characteristics which must be defined. Cell strains are cells that have been adapted to culture but, unlike cell lines, have a finite division potential. Non-immortalized cells stop dividing after 40 to 60 population doublings and, after this, they lose their ability to proliferate (a genetically determined event known as senescence).

Applications of cell culture

Mass culture of animal cell lines is fundamental to the manufacture of viral vaccines and other products of biotechnology. Culture of human stem cells is used to expand the number of cells and differentiate the cells into various somatic cell types for transplantation. Stem cell culture is also used to harvest the molecules and exosomes that the stem cells release for the purposes of therapeutic development.

Biological products produced by recombinant DNA (rDNA) technology in animal cell cultures include enzymes, synthetic hormones, immunobiologicals (monoclonal antibodies, interleukins, lymphokines), and anticancer agents. Although many simpler proteins can be produced using rDNA in bacterial cultures, more complex proteins that are glycosylated (carbohydrate-modified) currently must be made in animal cells. An important example of such a complex protein is the hormone erythropoietin. The cost of growing mammalian cell cultures is high, so research is underway to produce such complex proteins in insect cells or in higher plants, use of single embryonic cell and somatic embryos as a source for direct gene transfer via particle bombardment, transit gene expression and confocal microscopy observation is one of its applications. It also offers to confirm single cell origin of somatic embryos and the asymmetry of the first cell division, which starts the process.

Cell Culture in Two Dimensions

Research in tissue engineering, stem cells and molecular biology primarily involves cultures of cells on flat plastic dishes. This technique is known as two-dimensional (2D) cell culture, and was first developed by Wilhelm Roux who, in 1885, removed a portion of the medullary plate of an embryonic chicken and maintained it in warm saline for several days on a flat glass plate. From the advance of polymer technology arose today's standard plastic dish for 2D cell culture, commonly known as the Petri dish. Julius Richard Petri, a German bacteriologist, is generally credited with this invention while working as an assistant to Robert Koch. Various researchers today also utilize culturing laboratory flasks, conicals, and even disposable bags like those used in single-use bioreactors.

Aside from Petri dishes, scientists have long been growing cells within biologically derived matrices such as collagen or fibrin, and more recently, on synthetic hydrogels such as polyacrylamide or PEG. They do this in order to elicit phenotypes that are not expressed on conventionally rigid substrates. There is growing interest in controlling matrix stiffness, a concept that has led to discoveries in fields such as:

- Stem cell self-renewal
- Lineage specification
- Cancer cell phenotype
- Fibrosis
- Hepatocyte function
- Mechanosensing

Cell Culture in Three Dimensions

Cell culture in three dimensions has been touted as "Biology's New Dimension". Nevertheless, the practice of cell culture remains based on varying combinations of single or multiple cell structures in 2D. That being said, there is an increase in use of 3D cell cultures in research areas including drug discovery, cancer biology, regenerative medicine and basic life science research. There are a variety of platforms used to facilitate the growth of three-dimensional cellular structures including scaffold systems such as hydrogel matrices and solid scaffolds, and scaffold-free systems such as low-adhesion plates, nanoparticle facilitated magnetic levitation, and hanging drop plates.

3D cell Culture in Hydrogels

As the natural extracellular matrix (ECM) is important in the survival, proliferation, differentiation and migration of the cells, different hydrogel culture matrices mimicking natural ECM structure are considered as potential approaches towards in vivo –like cell culturing. Hydrogels are composed of interconnected pores with high water retention, which enables efficient transport of e.g. nutrients and gases. Several different types of hydrogels from natural and synthetic materials are available for 3D cell culture, including e.g. animal ECM extract hydrogels, protein hydrogels, peptide hydrogels, polymer hydrogels, and wood-based nanocellulose hydrogel.

3D Cell Culturing by Magnetic Levitation

3D Cell Culturing by Magnetic Levitation method (MLM) is the application of growing 3D tissue by inducing cells treated with magnetic nanoparticle assemblies in spatially varying magnetic fields using neodymium magnetic drivers and promoting cell to cell interactions by levitating the cells up to the air/liquid interface of a standard petri dish. The magnetic nanoparticle assemblies consist of magnetic iron oxide nanoparticles, gold nanoparticles, and the polymer polylysine. 3D cell culturing is scalable, with the capability for culturing 500 cells to millions of cells or from single dish to high-throughput low volume systems.

Tissue culture and Engineering

Cell culture is a fundamental component of tissue culture and tissue engineering, as it establishes the basics of growing and maintaining cells *in vitro*. The major application of human cell culture is in stem cell industry, where mesenchymal stem cells can be cultured and cryopreserved for future use. Tissue engineering potentially offers dramatic improvements in low cost medical care for hundreds of thousands of patients annually.

Vaccines

Vaccines for polio, measles, mumps, rubella, and chickenpox are currently made in cell cultures. Due to the H5N1 pandemic threat, research into using cell culture for influenza vaccines is being funded by the United States government. Novel ideas in the field include recombinant DNA-based vaccines, such as one made using human adenovirus (a common cold virus) as a vector, and novel adjuvants.

Culture of Non-Mammalian Cells

Plant Cell Culture Methods

Plant cell cultures are typically grown as cell suspension cultures in a liquid medium or as callus cultures on a solid medium. The culturing of undifferentiated plant cells and calli requires the proper balance of the plant growth hormones auxin and cytokinin.

Insect Cell Culture

Cells derived from Drosophila melanogaster (most prominently, Schneider 2 cells) can be used for experiments which may be hard to do on live flies or larvae, such as biochemical studies or studies using siRNA. Cell lines derived from the army worm *Spodoptera frugiperda*, including Sf9 and Sf21, and from the cabbage looper *Trichoplusia ni*, High Five cells, are commonly used for expression of recombinant proteins using baculovirus.

Bacterial and Yeast Culture Methods

For bacteria and yeasts, small quantities of cells are usually grown on a solid support that contains nutrients embedded in it, usually a gel such as agar, while large-scale cultures are grown with the cells suspended in a nutrient broth.

Viral Culture Methods

The culture of viruses requires the culture of cells of mammalian, plant, fungal or bacterial origin as hosts for the growth and replication of the virus. Whole wild type viruses, recombinant viruses or viral products may be generated in cell types other than their natural hosts under the right conditions. Depending on the species of the virus, infection and viral replication may result in host cell lysis and formation of a viral plaque.

Common Cell Lines

Human cell lines

- DU145 (prostate cancer)
- H295R (adrenocortical cancer)
- HeLa (cervical cancer)
- KBM-7 (chronic myelogenous leukemia)
- LNCaP (prostate cancer)
- MCF-7 (breast cancer)
- MDA-MB-468 (breast cancer)
- PC3 (prostate cancer)
- SaOS-2 (bone cancer)
- SH-SY5Y (neuroblastoma, cloned from a myeloma)
- T47D (breast cancer)
- THP-1 (acute myeloid leukemia)
- U87 (glioblastoma)
- National Cancer Institute's 60 cancer cell line panel (NCI60)

Primate cell lines

- Vero (African green monkey *Chlorocebus* kidney epithelial cell line)

Mouse cell lines

- MC3T3 (embryonic calvarium)

Rat tumor cell lines

- GH3 (pituitary tumor)
- PC12 (pheochromocytoma)

Plant cell lines

- Tobacco BY-2 cells (kept as cell suspension culture, they are model system of plant cell)

Other species cell lines

- Dog MDCK kidney epithelial
- Xenopus A6 kidney epithelial
- Zebrafish AB9

List of cell lines

This list is incomplete; you can help by expanding it.

Cell line	Meaning	Organism	Origin tissue	Morphology	Links
3T3-L1	"3-day transfer, inoculum 3 x 10^5 cells"	Mouse	Embryo	Fibroblast	ECACC Cellosaurus
4T1		Mouse	Mammary gland		ATCC Cellosaurus
9L		Rat	Brain	Glioblastoma	ECACC Cellosaurus
A172		Human	Brain	Glioblastoma	ECACC Cellosaurus
A20		Mouse	B lymphoma	B lymphocyte	Cellosaurus
A253		Human	Submandibular duct	Head and neck carcinoma	ATCC Cellosaurus
A2780		Human	Ovary	Ovarian carcinoma	ECACC Cellosaurus
A2780ADR		Human	Ovary	Adriamycin-resistant derivative of A2780	ECACC Cellosaurus
A2780cis		Human	Ovary	Cisplatin-resistant derivative of A2780	ECACC Cellosaurus
A431		Human	Skin epithelium	Squamous cell carcinoma	ECACC Cellosaurus
A549		Human	Lung	Lung carcinoma	ECACC Cellosaurus
AB9		Zebrafish	Fin	Fibroblast	ATCC Cellosaurus

Cell line	Meaning	Organism	Origin tissue	Morphology	Links
AHL-1	Armenian Hamster Lung-1	Hamster	Lung		ECACC Cellosaurus
ALC		Mouse	Bone marrow	Stroma	PMID 2435412 Cellosaurus
B16		Mouse	Melanoma		ECACC Cellosaurus
B35		Rat	Neuroblastoma		ATCC Cellosaurus
BCP-1		Human	PBMC	HIV+ primary effusion lymphoma	ATCC Cellosaurus
BEAS-2B	Bronchial epithelium + Adenovirus 12-SV40 virus hybrid (Ad12SV40)	Human	Lung	Epithelial	ECACC Cellosaurus
bEnd.3	Brain Endothelial 3	Mouse	Brain/cerebral cortex	Endothelium	Cellosaurus
BHK-21	Baby Hamster Kidney-21	Hamster	Kidney	Fibroblast	ECACC Cellosaurus
BOSC23	Packaging cell line derived from HEK 293	Human	Kidney (embryonic)	Epithelium	Cellosaurus
BT-20	Breast Tumor-20	Human	Breast epithelium	Breast carcinoma	ATCC Cellosaurus
BxPC3	Biopsy xenograft of Pancreatic Carcinoma line 3	Human	Pancreatic adenocarcinoma	Epithelial	ECACC Cellosaurus
C2C12		Mouse	Myoblast		ECACC Cellosaurus
C3H-10T1/2		Mouse	Embryonic mesenchymal cell line		ECACC Cellosaurus
C6		Rat	Brain astrocyte	Glioma	ECACC Cellosaurus
C6/36		Insect - Asian tiger mosquito	Larval tissue		ECACC Cellosaurus
Caco-2		Human	Colon	Colorectal carcinoma	ECACC Cellosaurus
Cal-27		Human	Tongue	Squamous cell carcinoma	ATCC Cellosaurus
CGR8		Mouse	Embryonic stem cells		ECACC Cellosaurus
CHO	Chinese Hamster Ovary	Hamster	Ovary	Epithelium	ECACC Cellosaurus
CML T1	Chronic myeloid leukemia T lymphocyte 1	Human	CML acute phase	T cell leukemia	DSMZ Cellosaurus

Cell line	Meaning	Organism	Origin tissue	Morphology	Links
CMT12	Canine Mammary Tumor 12	Dog	Mammary gland	Epithelium	Cellosaurus
COR-L23		Human	Lung	Lung carcinoma	ECACC Cellosaurus
COR-L23/5010		Human	Lung	Lung carcinoma	ECACC Cellosaurus
COR-L23/CPR		Human	Lung	Lung carcinoma	ECACC Cellosaurus
COR-L23/R23-		Human	Lung	Lung carcinoma	ECACC Cellosaurus
COS-7	*Cercopithecus aethiops*, origin-defective SV-40	Old World monkey - *Cercopithecus aethiops* (*Chlorocebus*)	Kidney	Fibroblast	ECACC Cellosaurus
COV-434		Human	Ovary	Ovarian granulosa cell carcinoma	PMID 8436435 ECACC Cellosaurus
CT26		Mouse	Colon	Colorectal carcinoma	Cellosaurus
D17		Dog	Lung metastasis	Osteosarcoma	ATCC Cellosaurus
DH82		Dog	Histiocytosis	Monocyte/macrophage	ECACC Cellosaurus
DU145		Human	Androgen insensitive prostate carcinoma		ATCC Cellosaurus
DuCaP	Dura mater cancer of the Prostate	Human	Metastatic prostate carcinoma	Epithelial	PMID 11317521 Cellosaurus
E14Tg2a		Mouse		Embryonic stem cells	ECACC Cellosaurus
EL4		Mouse		T cell leukemia	ECACC Cellosaurus
EM-2		Human	CML blast crisis	Ph+ CML line	DSMZ Cellosaurus
EM-3		Human	CML blast crisis	Ph+ CML line	DSMZ Cellosaurus
EMT6/AR1		Mouse	Mammary gland	Epithelial-like	ECACC Cellosaurus
EMT6/AR10.0		Mouse	Mammary gland	Epithelial-like	ECACC Cellosaurus
FM3		Human	Lymph node metastasis	Melanoma	ECACC Cellosaurus
GL261	Glioma 261	Mouse	Brain	Glioma	Cellosaurus
H1299		Human	Lung	Lung carcinoma	ATCC Cellosaurus

Cell line	Meaning	Organism	Origin tissue	Morphology	Links
HaCaT		Human	Skin	Keratinocyte	CLS Cellosaurus
HCA2		Human	Colon	Adenocarcinoma	ECACC Cellosaurus
HEK 293	Human Embryonic Kidney 293	Human	Kidney (embryonic)	Epithelium	ECACC Cellosaurus
HEK 293T	HEK 293 derivative	Human	Kidney (embryonic)	Epithelium	ECACC Cellosaurus
HeLa	"Henrietta Lacks"	Human	Cervix epithelium	Cervical carcinoma	ECACC Cellosaurus
Hepa1c1c7	Clone 7 of clone 1 hepatoma line 1	Mouse	Hepatoma	Epithelial	ECACC Cellosaurus
Hep G2		Human	Liver	Hepatoblastoma	ECACC Cellosaurus
High Five		Insect (moth) - *Trichoplusia ni*	Ovary		Cellosaurus
HL-60	Human Leukemia-60	Human	Blood	Myeloblast	ECACC Cellosaurus
HT-1080		Human		Fibrosarcoma	ECACC Cellosaurus
HT-29		Human	Colon epithelium	Adenocarcinoma	ECACC Cellosaurus
J558L		Mouse	Myeloma	B lymphocyte cell	ECACC Cellosaurus
Jurkat		Human	White blood cells	T cell leukemia	ECACC Cellosaurus
JY		Human	Lymphoblastoid	EBV-transformed B cell	ECACC Cellosaurus
K562		Human	Lymphoblastoid	CML blast crisis	ECACC Cellosaurus
KBM-7		Human	Lymphoblastoid	CML blast crisis	Cellosaurus
KCL-22		Human	Lymphoblastoid	CML	DSMZ Cellosaurus
KG1		Human	Lymphoblastoid	AML	ECACC Cellosaurus
Ku812		Human	Lymphoblastoid	Erythroleukemia	ECACC Cellosaurus
KYO-1	Kyoto-1	Human	Lymphoblastoid	CML	DSMZ Cellosaurus
L1210		Mouse	Lymphocytic leukemia	Ascitic fluid	ECACC Cellosaurus
L243		Mouse	Hybridoma	Secretes L243 mAb (against HLA-DR)	ATCC Cellosaurus
LNCaP	Lymph Node Cancer of the Prostate	Human	Prostatic adenocarcinoma	Epithelial	ECACC Cellosaurus

Cell line	Meaning	Organism	Origin tissue	Morphology	Links
MA2.1		Mouse	Hybridoma	Secretes MA2.1 mAb (against HLA-A2 and HLA-B17)	ATCC Cellosaurus
Ma-Mel 1, 2, 3....48		Human	Skin	A range of melanoma cell lines	ECACC Cellosaurus
MC-38	Mouse Colon-38	Mouse	Colon	Adenocarcinoma	Cellosaurus
MCF-7	Michigan Cancer Foundation-7	Human	Breast	Invasive breast ductal carcinoma ER+, PR+	ECACC Cellosaurus
MCF-10A	Michigan Cancer Foundation-10A	Human	Breast epithelium		ATCC Cellosaurus
MDA-MB-157	M.D. Anderson - Metastatic Breast-157	Human	Pleural effusion metastasis	Breast carcinoma	ECACC Cellosaurus
MDA-MB-231	M.D. Anderson - Metastatic Breast-231	Human	Pleural effusion metastasis	Breast carcinoma	ECACC Cellosaurus
MDA-MB-361	M.D. Anderson - Metastatic Breast-361	Human		Melanoma (contaminated by M14)	ECACC Cellosaurus
MDA-MB-468	M.D. Anderson - Metastatic Breast-468	Human	Pleural effusion metastasis	Breast carcinoma	ATCC Cellosaurus
MDCK II	Madin Darby Canine Kidney II	Dog	Kidney	Epithelium	ECACC Cellosaurus
MG63		Human	Bone	Osteosarcoma	ECACC Cellosaurus
MOR/0.2R		Human	Lung	Lung carcinoma	ECACC Cellosaurus
Mono-Mac-6		Human	White blood cells	Myeloid metaplasic AML	DSMZ Cellosaurus
MRC-5	Medical Research Council cell strain 5	Human	Lung (fetal)	Fibroblast	ECACC Cellosaurus
MTD-1A		Mouse		Epithelium	Cellosaurus
MyEnd	Myocardial Endothelial	Mouse		Endothelium	Cellosaurus
NCI-H69		Human	Lung	Lung carcinoma	ECACC Cellosaurus
NCI-H69/CPR		Human	Lung	Lung carcinoma	ECACC Cellosaurus
NCI-H69/LX10		Human	Lung	Lung carcinoma	ECACC Cellosaurus
NCI-H69/LX20		Human	Lung	Lung carcinoma	ECACC Cellosaurus
NCI-H69/LX4		Human	Lung	Lung carcinoma	ECACC Cellosaurus

Cell line	Meaning	Organism	Origin tissue	Morphology	Links
Neuro-2a		Mouse	Nerve/ neuroblastoma	Neuronal stem cells	ECACC Cellosaurus
NIH-3T3	NIH, 3-day transfer, inoculum 3 x 10^5 cells	Mouse	Embryo	Fibroblast	ECACC Cellosaurus
NALM-1		Human	Peripheral blood	Blast-crisis CML	ATCC Cellosaurus
Neuro2a		Mouse	Nerve/ neuroblastoma	Neuronal stem cells	Cellosaurus
NK-92		Human	Leukemia/ lymphoma		ATCC Cellosaurus
NTERA-2		Human	Lung metastasis	Embryonal carcinoma	ECACC Cellosaurus
NW-145		Human	Skin	Melanoma	ESTDAB Cellosaurus
OK	Opossum Kidney	Virginia opossum - *Didelphis virginiana*	Kidney		ECACC Cellosaurus
OPCN / OPCT cell lines		Human	Prostate	Range of prostate tumour lines	Cellosaurus
P3X63Ag8		Mouse	Myeloma		ECACC Cellosaurus
PC12		Rat	Adrenal medulla	Pheochromocytoma	ECACC Cellosaurus
PC-3	Prostate Cancer-3	Human	Bone metastasis	Prostate carcinoma	ECACC Cellosaurus
Peer		Human	T cell leukemia		DSMZ Cellosaurus
PNT1A		Human	Prostate	SV40-transformed tumour line	ECACC Cellosaurus
PNT2		Human	Prostate	SV40-transformed tumour line	ECACC Cellosaurus
Pt K2	The second cell line derived from *Potorous tridactylis*	Long-nosed potoroo - *Potorous tridactylus*	Kidney	Epithelial	ECACC Cellosaurus
Raji		Human	B lymphoma	Lymphoblast-like	ECACC Cellosaurus
RBL-1	Rat Basophilic Leukemia-1	Rat	Leukemia	Basophil cell	ECACC Cellosaurus
RenCa	Renal Carcinoma	Mouse	Kidney	Renal carcinoma	ATCC Cellosaurus
RIN-5F		Mouse	Pancreas		ECACC Cellosaurus
RMA-S		Mouse		T cell tumour	Cellosaurus

Cell line	Meaning	Organism	Origin tissue	Morphology	Links
S2	Schneider 2	Insect - *Drosophila melanogaster*	Late stage (20–24 hours old) embryos		ATCC Cellosaurus
SaOS-2	Sarcoma OSteogenic-2	Human	Bone	Osteosarcoma	ECACC Cellosaurus
Sf21	*Spodoptera frugiperda* 21	Insect (moth) - *Spodoptera frugiperda*	Ovary		ECACC Cellosaurus
Sf9	*Spodoptera frugiperda* 9	Insect (moth) - *Spodoptera frugiperda*	Ovary		ECACC Cellosaurus
SH-SY5Y		Human	Bone marrow metastasis	Neuroblastoma	ECACC Cellosaurus
SiHa		Human	Cervix epithelium	Cervical carcinoma	ATCC Cellosaurus
SK-BR-3	Sloan-Kettering Breast cancer 3	Human	Breast	Breast carcinoma	DSMZ Cellosaurus
SK-OV-3	Sloan-Kettering Ovarian cancer 3	Human	Ovary	Ovarian carcinoma	ECACC Cellosaurus
T2		Human		T cell leukemia/B cell line hybridoma	ATCC Cellosaurus
T-47D		Human	Breast	Breast ductal carcinoma	ECACC Cellosaurus
T84		Human	Lung metastasis	Colorectal carcinoma	ECACC Cellosaurus
T98G		Human	Glioblastoma-astrocytoma	Epithelium	ECACC Cellosaurus
THP-1		Human	Monocyte	Acute monocytic leukemia	ECACC Cellosaurus
U373		Human	Glioblastoma-astrocytoma	Epithelium	ECACC Cellosaurus
U87		Human	Glioblastoma-astrocytoma	Epithelial-like	ECACC Cellosaurus
U937		Human	Leukemic monocytic lymphoma		ECACC Cellosaurus
VCaP	Vertebral Cancer of the Prostate	Human	Vertebra metastasis	Prostate carcinoma	ECACC Cellosaurus
Vero	*Vero* (truth)	African green monkey - *Chlorocebus sabaeus*	Kidney epithelium		ECACC Cellosaurus
VG-1		Human		Primary effusion lymphoma	Cellosaurus
WM39		Human	Skin	Melanoma	ESTDAB Cellosaurus
WT-49		Human	Lymphoblastoid		ECACC Cellosaurus

Cell line	Meaning	Organism	Origin tissue	Morphology	Links
YAC-1		Mouse	Lymphoma		ECACC Cellosaurus
YAR		Human	Lymphoblastoid	EBV-transformed B cell	Human Immunology ECACC Cellosaurus

Autophagy

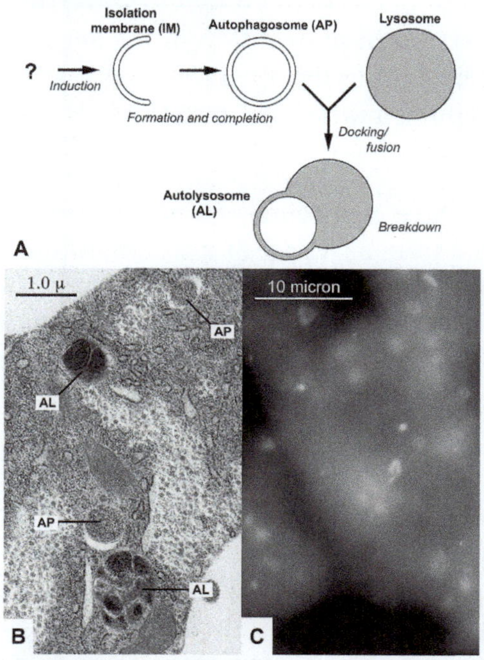

(A) Diagram of autophagy; (B) Electron micrograph of autophagic structures in the fatbody of a fruit fly larva; (C) Fluorescently labeled autophagosomes in liver cells of starved mice.

Autophagy (or autophagocytosis) (from the Greek *auto-*, "self" and *phagein*, "to eat"), is the natural, destructive mechanism that disassembles, through a regulated process, unnecessary or dysfunctional cellular components.

Autophagy allows the orderly degradation and recycling of cellular components. During this process, targeted cytoplasmic constituents are isolated from the rest of the cell within a double-membraned vesicle known as an autophagosome. The autophagosome then fuses with a lysosome and the contents are degraded and recycled. Three different forms of autophagy are commonly described: macroautophagy, microautophagy and chaperone-mediated autophagy. In the context of disease, autophagy has been seen as an adaptive response to stress, which promotes survival, whereas in other cases it appears to promote cell death and morbidity. In the extreme case of starvation, the breakdown of cellular components promotes cellular survival by maintaining cellular energy levels.

The name "autophagy" was coined by Belgian biochemist Christian de Duve in 1963.

History

Autophagy was first observed by Keith R. Porter and his student Thomas Ashford at the Rockefeller Institute. In January 1962 they reported an increased number of lysosomes in rat liver cells after addition of glucagon, and that some displaced lysosomes towards the centre of the cell contained other cell organelles such as mitochondria. They called this autolysis after Christian de Duve. However Porter and Ashford wrongly interpreted their data as lysosome formation (ignoring the pre-existing organelles). Lysosomes could not be cell organelles, but part of cytoplasm such as mitochondria, and that hydrolytic enzymes were produced by microbodies.

In 1963 researchers published a detailed ultrastructural description of "focal cytoplasmic degradation," which referenced a 1955 German study of injury-induced sequestration. The study recognized three continuous stages of maturation of the sequestered cytoplasm to lysosomes, and that the process was not limited to injury states that functioned under physiological conditions for "reutilization of cellular materials," and the "disposal of organelles" during differentiation.

Inspired by this discovery, de Duve christened the phenomena "autophagy". Unlike Porter and Ashford, de Duve conceived the term as a part of lysosomal function while describing the role of glucagon as a major inducer of cell degradation in the liver. With his student Peter, he established that lysosomes are responsible for glucagon-induced autophagy. This was the first time the fact that lysosomes were established as the sites of intracellular autophagy.

Process and Pathways

There are three pathways of autophagy and these are mediated by the autophagy-related genes and their associated enzymes.

Macroautophagy is the main pathway, used primarily to eradicate damaged cell organelles or unused proteins. This involves the formation of a double membrane known as an autophagosome, about the organelle marked for destruction. The autophagosome then travels through the cytoplasm of the cell to a lysosome, and the two organelles fuse. Within the lysosome, the contents of the autophagosome are degraded via acidic lysosomal hydrolases.

Microautophagy, on the other hand, involves the direct engulfment of cytoplasmic material into the lysosome. This occurs by invagination, meaning the inward folding of the lysosomal membrane, or cellular protrusion.

Chaperone-mediated autophagy, or CMA, is a very complex and specific pathway, which involves the recognition by the hsc70-containing complex. This means that a protein must contain the recognition site for this hsc70 complex which will allow it to bind to this chaperone, forming the CMA- substrate/chaperone complex. This complex then moves to the lysosomal membrane-bound protein that will recognise and bind with the CMA receptor, allowing it to enter the cell. Upon recognition, the substrate protein gets unfolded and it is translocated across the lysosome membrane with the assistance of the lysosomal hsc70 chaperone. CMA is significantly different from other types of autophagy because it translocates protein material in a one by one manner, and it is extremely selective about what material crosses the lysosomal barrier.

Molecular Biology

Autophagy is executed by autophagy-related (Atg) genes. The first autophagy genes were identified by genetic screens conducted in the budding yeast Saccharomyces cerevisiae. Following their identification those genes were functionally characterized and their orthologs in a variety of different organisms were identified and studied. In mammals, amino acid sensing and additional signals such as growth factors and reactive oxygen species regulate the activity of the protein kinases mTOR and AMPK. These two kinases regulate autophagy through inhibitory phosphorylation of the Unc-51-like kinases ULK1 and ULK2 (mammalian homologues of Atg1). Induction of autophagy results in the dephosphorylation and activation of the ULK kinases. ULK is part of a protein complex containing Atg13, Atg101 and FIP200. ULK phosphorylates and activates Beclin-1 (mammalian homologue of Atg6), which is also part of a protein complex. The autophagy-inducible Beclin-1 complex contains the proteins p150, Atg14L and the class III phosphatidylinositol 3-phosphate kinase (PI(3)K) VPS34. The active ULK and Beclin-1 complexes re-localize to the site of autophagosome initiation, the phagophore, where they both contribute to the activation of downstream autophagy components. Once active, VPS34 phosphorylates the lipid phosphatidylinositol to generate phosphatidylinositol 3-phosphate (PtdIns(3)P) on the surface of the phagophore. The generated PtdIns(3)P is used as a docking point for proteins harboring a PtdIns(3)P binding motif. WIPI2, a PtdIns(3) P binding protein of the WIPI (WD-repeat protein interacting with phosphoinositides) protein family, was recently shown to physically bind Atg16L1. Atg16L1 is a member of an E3-like protein complex involved in one of two ubiquitin-like conjugation systems essential for autophagosome formation. Its binding by WIPI2 recruits it to the phagophore and mediates its activity. The first of the two ubiquitin-like conjugation systems involved in autophagy covalently binds the ubiquitin-like protein Atg12 to Atg5. The resulting conjugate protein then binds Atg16L1 to form an E3-like complex which functions as part of the second ubiquitin-like conjugation system. This complex binds and activates Atg3, which covalently attaches mammalian homologues of the ubiquitin-like yeast protein Atg8 (LC3A-C, GATE16, and GABARAPL1-3), the most studied being LC3 proteins, to the lipid phosphatidylethanolamine (PE) on the surface of autophagosomes. Lipidated LC3 contributes to the closure of autophagosomes, and enables the docking of specific cargos and adaptor proteins such as p62. The completed autophagosome then fuses with a lysosome through the actions of multiple proteins, including SNAREs and UVRAG. Following the fusion LC3 is retained on the vesicle's inner side and degraded along with the cargo, while the LC3 molecules attached to the outer side are cleaved off by Atg4 and recycled. The contents of the autolysosome are subsequently degraded and their building blocks are released from the vesicle through the action of permeases.

Functions

Nutrient Starvation

Autophagy has roles in various cellular functions. One particular example is in yeasts, where the nutrient starvation induces a high level of autophagy. This allows unneeded proteins to be degraded and the amino acids recycled for the synthesis of proteins that are essential for survival. In higher eukaryotes, autophagy is induced in response to the nutrient depletion that occurs in animals at birth after severing of the trans-placental food supply, as well as that of nutrient starved cultured cells and tissues. Mutant yeast cells that have a reduced autophagic capability rapidly perish in nutrition-deficient conditions. Studies on the *apg* mutants suggest that autophagy via autophagic bodies is indispensable for protein degradation in the vacuoles under starvation conditions, and that at least 15 APG genes are involved in autophagy in yeast. A gene known as ATG7 has been implicated in nutrient-mediated autophagy, as mice studies have shown that starvation-induced autophagy was impaired in *atg7*-deficient mice.

Xenophagy

In microbiology, xenophagy is the autophagic degradation of infectious particles. Cellular autophagic machinery also play an important role in innate immunity. Intracellular pathogens, such as *Mycobacterium tuberculosis* (the bacterium which is responsible for tuberculosis) are targeted for degradation by the same cellular machinery and regulatory mechanisms that target host mitochondria for degradation. Incidentally, this is further evidence for the endosymbiotic hypothesis. This process generally leads to the destruction of the invasive organism, although some bacteria can block the maturation of phagosomes into degradative organelles called phagolysosomes. Stimulation of autophagy in infected cells can help overcome this phenomenon, restoring pathogen degradation.

Autophagy and Infection

Vesicular stomatitis virus is believed to be taken up by the autophagosome from the cytosol and translocated to the endosomes where detection takes place by a member of the PRRs called toll-like receptor 7, detecting single stranded RNA. Following activation of the toll-like receptor, intracellular signaling cascades are initiated, leading to induction of interferon and other antiviral cytokines. A subset of viruses and bacteria subvert the autophagic pathway to promote their own replication. Galectin-8 has recently been identified as an intracellular "danger receptor", able to initiate autophagy against intracellular pathogens. When galectin-8 binds to a damaged vacuole, it recruits autophagy adaptor such as NDP52 leading to the formation of an autophagosome and bacterial degradation.

Repair Mechanism

Autophagy degrades damaged organelles, cell membranes and proteins, and the failure of autophagy is thought to be one of the main reasons for the accumulation of cell damage and aging.

Programmed Cell Death

One of the mechanisms of programmed cell death (PCD) is associated with the appearance of autophagosomes and depends on autophagy proteins. This form of cell death most likely corresponds to a process that has been morphologically defined as autophagic PCD. One question that constantly arises, however, is whether autophagic activity in dying cells is the cause of death or is actually an attempt to prevent it. Morphological and histochemical studies so far did not prove a causative relationship between the autophagic process and cell death. In fact, there have recently been strong arguments that autophagic activity in dying cells might actually be a survival mechanism. Studies of the metamorphosis of insects have shown cells undergoing a form of PCD that appears distinct from other forms; these have been proposed as examples of autophagic cell death.

Recent pharmacological and biochemical studies have proposed that survival and lethal autophagy can be distinguished by the type and degree of regulatory signaling during stress particularly after viral infection. Although promising, these findings have not been examined in non-viral systems.

Autophagy and Caloric Restriction

Research suggests that autophagy is required for the lifespan-prolonging effects of caloric restriction. A 2010 French study of nematodes, mice and flies showed that inhibition of autophagy exposed cells to metabolic stress. Resveratrol and the dietary restriction prolonged the lifespan of normal, autophagy-proficient nematodes, but not of nematodes in which autophagy had been inhibited by knocking out Beclin 1 (a known autophagic modulator).

Autophagy and Exercise

Autophagy is essential for basal homeostasis; it is also extremely important in maintaining muscle homeostasis during physical exercise. Autophagy at the molecular level is only partially understood. A study of mice shows that autophagy is important for the ever changing demands of their nutritional and energy needs, particularly through the metabolic pathways of protein catabolism. In a 2012 study conducted by the University of Texas Southwestern Medical Center in Dallas, mutant mice (with a knock-in mutation of BCL2 phosphorylation sites to produce progeny that showed normal levels of basal autophagy yet were deficient in stress-induced autophagy) were tested to challenge this theory. Results showed that when compared to a control group, these mice illustrated a decrease in endurance and an altered glucose metabolism during acute exercise.

Another study demonstrated that skeletal muscle fibres of collagen VI knockout mice showed signs of degeneration due to an insufficiency of autophagy which led to an accumulation of damaged mitochondria and excessive cell death. Exercise-induced autophagy was unsuccessful however; but when autophagy was induced artificially post-exercise, the accumulation of damaged organelles in collagen VI deficient muscle fibres was prevented and cellular homeostasis was maintained. Both studies demonstrate that autophagy induction may contribute to the beneficial metabolic effects of exercise and that it is essential in the maintaining of muscle homeostasis during exercise, particularly in collagen VI fibres.

Work at the Institute for Cell Biology, University of Bonn, showed that a certain type of autophagy, i.e., chaperone-assisted selective autophagy (CASA), is induced in contracting muscles and is required for maintaining the muscle sarcomere under mechanical tension. The CASA chaperone complex recognizes mechanically damaged cytoskeleton components and directs these components through a ubiquitin-dependent autophagic sorting pathway to lysosomes for disposal. This is necessary for maintaining muscle activity.

Autophagy and Osteoarthritis

Because autophagy decreases with age and age is a major risk factor for osteoarthritis, the role of autophagy in the development of this disease is suggested. Proteins involved in autophagy are reduced with age in both human and mouse articular cartilage. Mechanical injury to cartilage explants in culture also reduced autophagy proteins. Autophagy is constantly activated in normal cartilage but it is compromised with age and precedes cartilage cell death and structural damage.

These results suggest autophagy is a normal protective process (chondroprotection) in the joint.

Autophagy and Cancer

Oftentimes, cancer occurs when several different pathways that regulate cell differentiation are disturbed. Autophagy plays an important role in cancer – both in protecting against cancer as well as potentially contributing to the growth of cancer. Autophagy can contribute to cancer by promoting survival of tumor cells that have been starved, or that degrade apoptotic mediators through autophagy: in such cases, use of inhibitors of the late stages of autophagy (such as chloroquine), on the cells that use autophagy to survive, increases the number of cancer cells killed by antineoplastic drugs.

The role of autophagy in cancer is one that has been highly researched and reviewed. There is evidence that emphasizes the role of autophagy both as a tumor suppressor as well as a factor in tumor cell survival. However, recent research has been able to show that autophagy is more likely to be used as a tumor suppressor according to several models.

Tumor Suppressor

Several experiments have been done with mice and varying Beclin1, a protein that regulates autophagy. When the Beclin1 gene was altered to be heterozygous (Beclin 1+/-), the mice were found to be tumor prone. However, when Beclin1 was overexpressed, tumor development was inhibited.

Necrosis and chronic inflammation also has been shown to be limited through autophagy which helps protect against the formation of tumor cells. Thus these experiments show autophagy's role as a tumor suppressor.

Tumor Cell Survival

Alternatively, autophagy has also been shown to play a huge role in tumor cell survival. In cancerous cells, autophagy is used as a way to deal with stress on the cell. Once these autophagy related genes were inhibited, cell death was potentiated. The increase in metabolic energy is offset by autophagy functions. These metabolic stresses include hypoxia, nutrient deprivation, and an increase in proliferation. These stresses activate autophagy in order to recycle ATP and maintain survival of the cancerous cells. Autophagy has been shown to enable continued growth of tumor cells by maintaining cellular energy production. By inhibiting autophagy genes in these tumors cells, regression of the tumor and extended survival of the organs affected by the tumors were found. Furthermore, inhibition of autophagy has also been shown to enhance the effectiveness of anticancer therapies.

Mechanism of Cell Death

Cells that undergo an extreme amount of stress experience cell death either through apoptosis or necrosis. Prolonged autophagy activation leads to a high turnover rate of proteins and organelles. A high rate above the survival threshold may kill cancer cells with a high apoptotic threshold. This technique can be utilized as a therapeutic cancer treatment.

Therapeutic Target

New developments in research have found that targeted autophagy may be a viable therapeutic solution in fighting cancer. As discussed above, autophagy plays both a role in tumor suppression and tumor cell survival. Thus, the qualities of autophagy can be used as a strategy for cancer prevention. The first strategy is to induce autophagy and enhance its tumor suppression attributes. The second strategy is to inhibit autophagy and thus induce apoptosis.

The first strategy has been tested by looking at dose-response anti-tumor effects during autophagy-induced therapies. These therapies have shown that autophagy increases in a dose-dependent manner. This is directly related to the growth of cancer cells in a dose-dependent manner as well. This data supports the development of therapies that will encourage autophagy. Secondly, inhibiting the protein pathways directly known to induce autophagy may also serve as an anticancer therapy.

The second strategy is based on the idea that autophagy is a protein degradation system used to maintain homeostasis and the findings that inhibition of autophagy often leads to apoptosis. Inhibition of autophagy is riskier as it may lead to cell survival instead of the desired cell death.

Mitophagy

Mitophagy is the selective degradation of mitochondria by autophagy. It often occurs to defective mitochondria following damage or stress. This process was first mentioned by J.J. Lemasters in 2005, although lysosomes in the liver that contained mitochondrial fragments had been seen as early as 1962, "As part of almost every lysosome in these glucagon-treated cells it is possible to recognize a mitochondrion or a remnant of one. It was also mentioned in 1977 by scientists studying metamorphosis in silkworms, "...mitochondria develop functional alterations which would activate autophagy." Mitophagy is key in keeping the cell healthy. It promotes turnover of mitochondria and prevents accumulation of dysfunctional mitochondria which can lead to cellular degeneration. It is mediated by Atg32 (in yeast) and NIP3-like protein X (NIX). Mitophagy is regulated by PINK1 and parkin protein. The occurrence of mitophagy is not limited to the damaged mitochondria but also involves undamaged ones.

Role

Organelles and bits of cytoplasm are sequestered and targeted for degradation by the lysosome for hydrolytic digestion by a process known as autophagy. Mitochondria metabolism leads to the creation of by-products that lead to DNA damage and mutations. Therefore, a healthy population of mitochondria is critical for the well-being of cells. Previously it was thought that targeted degradation of mitochondria was a stochastic event, but accumulating evidence suggest that mitophagy is a selective process.

Generation of ATP by oxidative phosphorylation leads to the production of various reactive oxygen species (ROS) in the mitochondria, and submitochondrial particles. Formation of ROS as a mitochondrial waste product will eventually lead to cytotoxicity and cell death. Because of their role in metabolism, mitochondria are very susceptible to ROS damage. Damaged mitochondria causes a depletion in ATP and a release of cytochrome *c*, which leads to activation of caspases and onset of apoptosis. Mitochondrial damage is not caused solely by oxidative stress or disease processes; normal mitochondria will eventually accumulate oxidative damage hallmarks overtime, which can be deleterious to mitochondria as well as to the cell. These faulty mitochondria can further deplete the cell from ATP, increase production of ROS, and release proapoptopic proteins such as caspases.

Because of the danger of having damaged mitochondria in the cell, the timely elimination of damaged and aged mitochondria is essential for maintaining the integrity of the cell. This turnover process consists of the sequestration and hydrolytic degradation by the lysosome, a process also known as mitophagy.

Mitochondrial depletion reduces a spectrum of senescence effectors and phenotypes while preserving ATP production via enhanced glycolysis.

Pathways

Mitophagy in Mammals

There are several ways that mitophagy is induced in mammalian cells. The PINK1 and Parkin pathway is, so far, the best characterized. This pathway starts in by deciphering the difference between healthy mitochondria and damaged mitochondria. A 64-kDa protein, PTEN-induced kinase 1 (PINK1), has been implicated to detect mitochondrial quality. PINK1 contains a mitochondrial targeting sequence (MTS) and is recruited to the mitochondria. In healthy mitochondria, PINK1 is imported through the outer membrane via the TOM complex, and partially through the inner mitochondrial membrane via the TIM complex, so it then spans the inner mitochondrial membrane. The process of import into the inner membrane is associated -----with the cleavage of PINK1 from 64-kDa into 60-kDa. Pink1 is then cleaved by PARL into 52-kDa. This new form of PINK1 is degraded by proteases within the mitochondria. This keeps the concentration of PINK1 in check in healthy mitochondria.

In unhealthy mitochondria, the interior mitochondrial membrane becomes depolarized. This membrane potential is necessary for the TIM-mediated protein import. In depolarized mitochondria, PINK1 is no longer imported into the inner membrane, is not cleaved by PARL and PINK1 concentration increases in the outer mitochondrial membrane. PINK1 can then recruit Parkin. It is thought that PINK1 phosphorylates Parkin ubiquitin at S65 which initiates Parkin recruitment at the mitochondria. Parkin is a cystolic E3 ubiquitin ligase 8. Once localized at the mitochondria, PINK1 phosphorylates Parkin at S65, homologous to the site where ubiquitin was phosphorylated, which activates Parkin by inducing dimerization and an active state. This allows for Parkin mediated ubiquitination on other proteins.

Because of the PINK1 mediated recruitment to the mitochondrial surface, Parkin can ubiquitylate proteins in the outer mitochondrial membrane. Some of these proteins include Mfn1/Mfn2 and mitoNEET. The ubiquitylation of mitochondrial surface proteins brings in mitophagy initiating factors. Parkin promotes ubiquitin chain linkages on both K63 and K48. K48 Ubiquitination initiates degradation of the proteins, and could allow for passive mitochondrial degradation. K63 ubiquitination is thought to recruit autophagy adaptors LC3/GABARAP which will then lead to mitophagy. It is still unclear which proteins are necessary and sufficient for mitophagy, and how these proteins, once ubiquitylated, initiate mitophagy.

Other pathways that can induce mitophagy includes mitophagy receptors on the outer mitochondrial membrane surface. These receptors include NIX1, BNIP3 and FUNDC1. All of these receptors contain LIR consensus sequences that bind LC3/GABARAP which can lead to the degradation of the mitochondria. In hypoxic conditions BNIP3 is upregulated by HIF1α. BNIP3 is then phosphorylated at its serine residues near the LIR sequence which promotes LC3 binding. FUNDCI is also hypoxia sensitive, although it is constitutively present at the outer mitochondrial membrane during normal conditions

Mitophagy in Yeast

Mitophagy in yeast was first presumed after the discovery of Yeast Mitochondrial Escape genes (yme), specifically yme1. Yme1 like other genes in the family showed increase escapes of mtDNA, but was the only one that showed in increase in mitochondrial degradation. Through work on this gene that mediates the escape of mtDNA, researchers discovered that mitochondrial turnover is triggered by proteins.

More was discovered about genetic control of mitophagy after studies done of UTH1. After doing a screen for genes that regulate longevity. It was found in ΔUTH1 strains there was an inhibition of mitophagy, which occurred without affecting autophagy mechanisms. It also showed that Uth1p protein is necessary to move mitochondria to the vacuole. This suggested there is a specialized system for mitophagy. Other studies looked at AUP1, a mitochondrial phosphatase, and found Aup1 marks mitochondria for elimination.

Another yeast protein associated with mitophagy is a mitochondrial inner membrane protein, Mdm38p/Mkh1p. This protein is part of the complex that exchanges K+/H+ ions across the inner membrane. Deletions to this protein causes swelling, a loss of membrane potential, and mitochondrial fragmentation.

Recently, it has been shown that ATG32 (autophagy related gene 32) plays a crucial role in yeast mitophagy. It is localized to the mitochondria. Once mitophagy is initiated, Atg32 binds to Atg11 and the Atg32-associated mitochondria is transported to the vacuole. Atg32 silencing stops recruitment of autophagy machinery and mitochondrial degradation. Atg32 is not necessary for other forms of autophagy.

All of these proteins likely play a role in maintaining a healthy mitochondria, but mutations have shown that dysregulation can lead to a selective degradation of mitochondria. Whether these proteins work in concert, are main players in mitophagy, or the networks controlled still remain to be elucidated

Relation to Disease

Mitophagy and Cancer

In 1920 Otto Warburg observed that certain cancerous tumors display a metabolic shift towards glycolysis. This hypothesis is referred to as the "Warburg Effect", in which cancer cells produce energy via the conversion of glucose into lactate, even in the presence of oxygen (aerobic glycolysis). Despite nearly a century since it was first described, a lot of questions remained unanswered regarding the Warburg effect. Initially, Warburg attributed this metabolic shift to mitochondrial dysfunction in cancer cells. Further studies in tumor biology have shown that the increased growth rate in cancer cells is due to an overdrive in glycolysis (glygolytic shift), which leads to a decrease in oxidative phosphorylation and mitochondrial density. As a consequence of the Warburg effect, cancer cells would produce large amounts of lactate. The excess lactate is then released to the extracellular environment which results in a decrease in extracellular pH. This micro-environment acidification can lead to cellular stress, which would lead to autophagy. Autophagy is activated in response to a range of stimuli, including nutrient depletion, hypoxia, and activated oncogenes. However, it appears that autophagy can help in cancer cell survival under conditions of metabolic stress and it may confer resistance to anti-cancer therapies such as radiation and chemotherapy. Additionally, in the microenvironment of cancer cells, there is an increase in hypoxia-inducible transcription factor 1-alpha (HIF1A), which promotes expression of BNIP3, an essential factor for mitophagy.

Mitophagy and Parkinson's disease

Parkinson's disease is a neurodegenerative disorder partially caused by the cell death of dopamine creating cells in the substantia nigra. There are several genetic mutations implicated in Parkinson's disease, including loss of function PINK1 and Parkin. Loss of function in these genes can lead to damaged mitochondrial accumulation and protein aggregates than can lead to cellular degeneration.

Mitochondria is thought to be involved in Parkinson's disease. In spontaneous, usually aging related Parkinson's disease (non-genetically linked), the disease is commonly caused by dysfunctional mitochondria, cellular oxidative stress, autophagic alterations and the aggregation of proteins. These can lead to mitochondrial swelling and depolarization. It is important to keep the dysfunctional mitochondria regulated, because all of these traits could be induced by mitochondrial dysfunction and can induce cell death. Disorders in energy creation by mitochondria can cause cellular degeneration, like those seen in the substantia nigra.

References

- Inwood, Stephen (2003). The man who knew too much: the strange and inventive life of Robert Hooke, 1635–1703. London: Pan. p. 72. ISBN 0-330-48829-5.
- Becker, Wayne M.; Kleinsmith, Lewis J.; Hardin, Jeff (2003). The World of the Cell. Benjamin/Cummings Publishing Company. p. 1. ISBN 978-0-8053-4854-5.
- William A. Dunn Jr.; Laura A. Schroder; John P. Aris (2013). "Historical overview of autophagy". In Hong-Gang Wang. Autophagy and Cancer. Springer. pp. 3–4. ISBN 9781461465614.
- Tavassoly, Iman (2015). Dynamics of Cell Fate Decision Mediated by the Interplay of Autophagy and Apoptosis in Cancer Cells. Springer International Publishing. ISBN 978-3-319-14962-2.
- Cuervo AM, Bergamini E, Brunk UT, Dröge W, Ffrench M, Terman A (2005). "Autophagy and aging: the importance of maintaining "clean" cells". Autophagy 1 (3): 131–40. doi:10.4161/auto.1.3.2017. PMID 16874025.
- Kobayashi S (2015). "Choose Delicately and Reuse Adequately: The Newly Revealed Process of Autophagy". Biological & Pharmaceutical Bulletin 38 (8): 1098–103. doi:10.1248/bpb.b15-00096. PMID 26235572.

- T. Proikas-Cezanne, Z. Takacs, P. Donnes, and O. Kohlbacher, 'Wipi Proteins: Essential Ptdins3p Effectors at the Nascent Autophagosome', J Cell Sci, 128 (2015), 207-17
- Tavassoly, Iman (2015). Dynamics of Cell Fate Decision Mediated by the Interplay of Autophagy and Apoptosis in Cancer Cells. Springer International Publishing. ISBN 978-3-319-14962-2.
- Carames, B; Olmer, M; Kiosses, WB; Lotz, MK (2015). "The relationship of autophagy defects to Cartilage Damage During joint aging in a mouse model". Arthritis Rheumatol 67 (6): 1568–1576. doi:10.1002/art.39073. PMID 25708836.
- Kalenderblatt Dezember 2013 – Mathematisch-Naturwissenschaftliche Fakultät – Universität Rostock. Math-nat.uni-rostock.de (2013-11-28). Retrieved on 2015-10-15.
- H. C. Dooley, M. Razi, H. E. Polson, S. E. Girardin, M. I. Wilson, and S. A. Tooze, 'Wipi2 Links Lc3 Conjugation with Pi3p, Autophagosome Formation, and Pathogen Clearance by Recruiting Atg12-5-16l1', Mol Cell, 55 (2014), 238-52
- S. Park, S. G. Choi, S. M. Yoo, J. H. Son, and Y. K. Jung, 'Choline Dehydrogenase Interacts with Sqstm1/P62 to Recruit Lc3 and Stimulate Mitophagy', Autophagy, 10 (2014), 1906-20
- Y. M. Kim, C. H. Jung, M. Seo, E. K. Kim, J. M. Park, S. S. Bae, and D. H. Kim, 'Mtorc1 Phosphorylates Uvrag to Negatively Regulate Autophagosome and Endosome Maturation', Mol Cell (2014)
- Baxevanis CN, Chrousos GP (2014). "New use for old drugs? Prospective targets of chloroquines in cancer therapy". Curr Drug Targets. 15 (9): 843–51. doi:10.2174/1389450115666140714121514. PMID 25023646.
- Derosa, Maria C.; Monreal, Carlos; Schnitzer, Morris; Walsh, Ryan; Sultan, Yasir (2010). "Nanotechnology in fertilizers". Nature Nanotechnology. 5 (2): 91. doi:10.1038/nnano.2010.2. PMID 20130583.
- C. A. Lamb, T. Yoshimori, and S. A. Tooze, 'The Autophagosome: Origins Unknown, Biogenesis Complex', Nat Rev Mol Cell Biol, 14 (2013), 759-74

5

Cell Cycle and Cell Reproduction

Cells originate and reproduce through cell division. All cells reach a stage where the cell divides into daughter cells. This is an important area of study in cell biology. This chapter will provide an integrated understanding of cell reproduction.

Cell Cycle

The cell cycle or cell-division cycle is the series of events that take place in a cell leading to its division and duplication of its DNA (DNA replication) to produce two daughter cells. In bacteria, which lack a cell nucleus, the cell cycle is divided into the B, C, and D periods. The B period extends from the end of cell division to the beginning of DNA replication. DNA replication occurs during the C period. The D period refers to the stage between the end of DNA replication and the splitting of the bacterial cell into two daughter cells. In cells with a nucleus, as in eukaryotes, the cell cycle is also divided into three periods: interphase, the mitotic (M) phase, and cytokinesis. During interphase, the cell grows, accumulating nutrients needed for mitosis, preparing it for cell division and duplicating its DNA. During the mitotic phase, the chromosomes separate. During the final stage, cytokinesis, the chromosomes and cytoplasm separate into two new daughter cells. To ensure the proper division of the cell, there are control mechanisms known as cell cycle checkpoints.

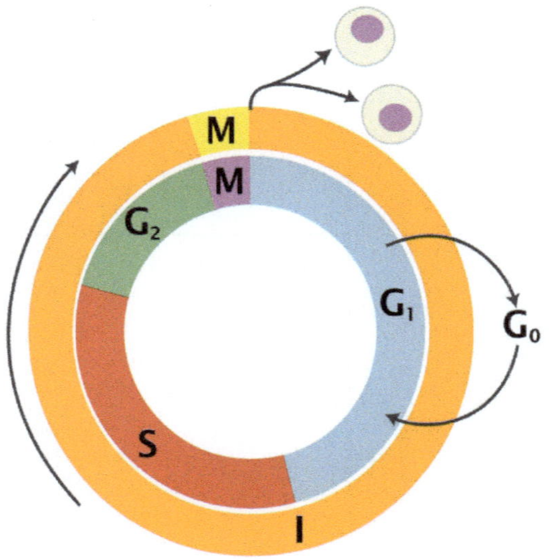

Schematic of the cell cycle. outer ring: I = Interphase, M = Mitosis; inner ring: M = Mitosis, G_1 = Gap 1, G_2 = Gap 2, S = Synthesis; not in ring: G_0 = Gap 0/Resting.

The cell-division cycle is a vital process by which a single-celled fertilized egg develops into a mature organism, as well as the process by which hair, skin, blood cells, and some internal organs are renewed. After cell division, each of the daughter cells begin the interphase of a new cycle. Although the various stages of interphase are not usually morphologically distinguishable, each phase of the cell cycle has a distinct set of specialized biochemical processes that prepare the cell for initiation of cell division.

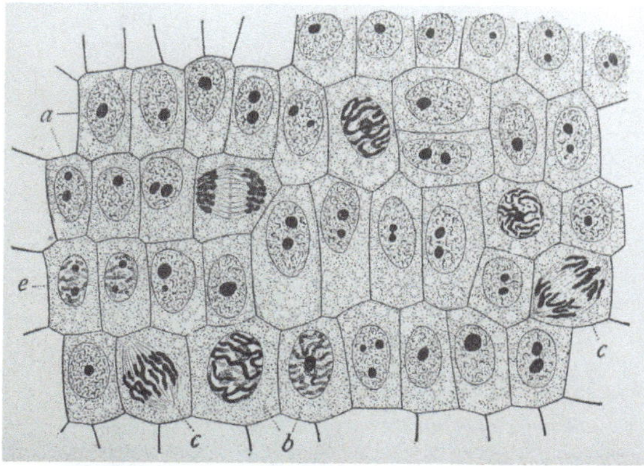

Onion (Allium) cells in different phases of the cell cycle. Growth in an organism is carefully controlled by regulating the cell cycle.

Cell Cycle Phases

State	Description	Abbreviation	:	
	Gap 0	G_0	A resting phase where the cell has left the cycle and has stopped dividing.Cell cycle starts with this phase.	
Quiescent senescent/ resting phase		Gap 1	G_1	Cells increase in size in Gap 1. The *G_1 checkpoint* control mechanism ensures that everything is ready for DNA synthesis.
	Interphase	Synthesis	**S**	DNA replication occurs during this phase.
Gap 2		G_2	During the gap between DNA synthesis and mitosis, the cell will continue to grow. The *G_2 checkpoint* control mechanism ensures that everything is ready to enter the M (mitosis) phase and divide.	
Cell division	Mitosis	**M**	Cell growth stops at this stage and cellular energy is focused on the orderly division into two daughter cells. A checkpoint in the middle of mitosis (*Metaphase Checkpoint*) ensures that the cell is ready to complete cell division.	

G_0 Phase (Quiescence)

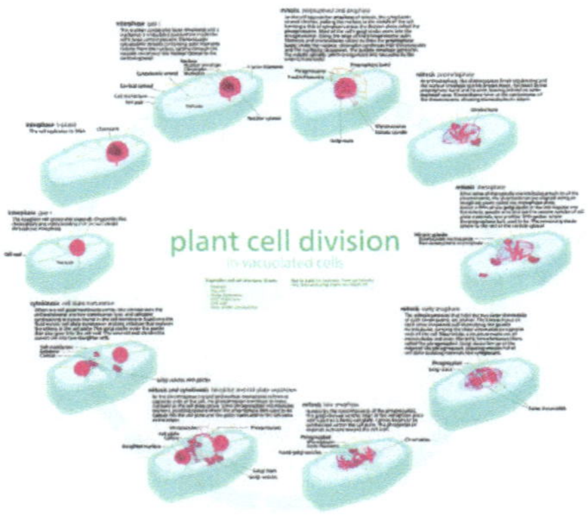

Plant cell cycle

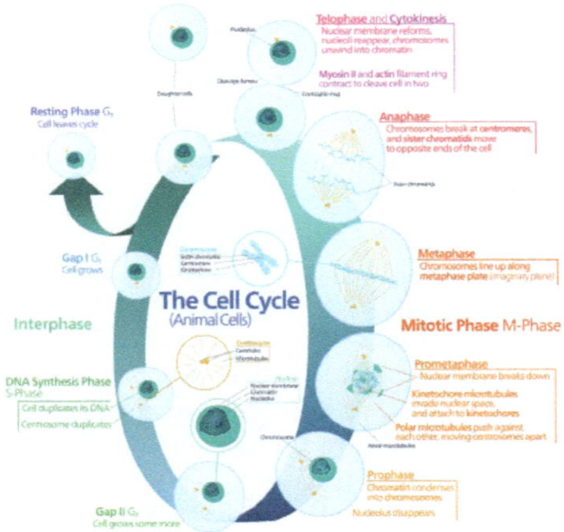

Animal cell cycle

The word "post-mitotic" is sometimes used to refer to both quiescent and senescent cells. Nonproliferative cells in multicellular eukaryotes generally enter the quiescent G_0 state from G_1 and may remain quiescent for long periods of time, possibly indefinitely (as is often the case for neurons). This is very common for cells that are fully differentiated. Cellular senescence occurs in response to DNA damage or degradation that would make a cell's progeny nonviable; for example, become cancerous. Some cells enter the G_0 phase semi-permanently e.g., some liver, kidney, stomach cells. Many cells do not enter G_0 and continue to divide throughout an organism's life, e.g. epithelial cells.

Interphase (Intermitosis)

Before a cell can enter cell division, it needs to take in nutrients. All of the preparations are done during interphase. Interphase is a series of changes that takes place in a newly formed cell and its nucleus, before it becomes capable of division again. It is also called preparatory phase or intermitosis. Previously it was called resting stage because there is no apparent activity related to cell division.Typically interphase lasts for at least 90% of the total time required for the cell cycle.

Interphase proceeds in three stages, G_1, S, and G_2, followed by the cycle of mitosis and cytokinesis. The cell's nuclear chromosomes are duplicated during S phase.

G_1 Phase (Growth)

The first phase within interphase, from the end of the previous M phase until the beginning of DNA synthesis, is called G_1 (G indicating *gap*). It is also called the growth phase. During this phase, the biosynthetic activities of the cell, which are considerably slowed down during M phase, resume at a high rate. The duration of G_1 is highly variable, even among different cells of the same species. In this phase, the cell increases its supply of proteins, increases the number of organelles (such as mitochondria, ribosomes), and grows in size.

S Phase (DNA Replication)

The ensuing S phase starts when DNA replication commences; when it is completed, all of the chromosomes have been replicated, i.e., each chromosome has two (sister) chromatids. Thus, during this phase, the amount of DNA in the cell has effectively doubled, though the ploidy of the cell remains the same. During this phase, synthesis is completed as quickly as possible due to the exposed base pairs being sensitive to harmful external factors such as mutagens

G_2 Phase (Growth)

G2 phase is a period of protein synthesis and rapid cell growth to prepare the cell for mitosis.

Mitotic Phase (Chromosome Separation)

The relatively brief *M phase* consists of nuclear division (karyokinesis). It is a relatively short period of the cell cycle. M phase is complex and highly regulated. The sequence of events is divided into phases, corresponding to the completion of one set of activities and the start of the next. These phases are sequentially known as:

- prophase,
- metaphase,
- anaphase,
- telophase

Mitosis is the process by which a eukaryotic cell separates the chromosomes in its cell nucleus into two identical sets in two nuclei. During the process of mitosis the pairs of chromosomes condense and attach to fibers that pull the sister chromatids to opposite sides of the cell.

Mitosis occurs exclusively in eukaryotic cells, but occurs in different ways in different species. For example, animals undergo an "open" mitosis, where the nuclear envelope breaks down before the chromosomes separate, while fungi such as *Aspergillus nidulans* and *Saccharomyces cerevisiae* (yeast) undergo a "closed" mitosis, where chromosomes divide within an intact cell nucleus. Prokaryotic cells, which lack a nucleus, divide by a process called binary fission.

Cytokinesis Phase (Separation of All Cell Components)

Mitosis is immediately followed by cytokinesis, which divides the nuclei, cytoplasm, organelles and cell membrane into two cells containing roughly equal shares of these cellular components.

Mitosis and cytokinesis together define the division of the mother cell into two daughter cells, genetically identical to each other and to their parent cell. This accounts for approximately 10% of the cell cycle.

Because cytokinesis usually occurs in conjunction with mitosis, "mitosis" is often used interchangeably with "M phase". However, there are many cells where mitosis and cytokinesis occur separately, forming single cells with multiple nuclei in a process called endoreplication. This occurs most notably among the fungi and slime moulds, but is found in various groups. Even in animals, cytokinesis and mitosis may occur independently, for instance during certain stages of fruit fly embryonic development. Errors in mitosis can either kill a cell through apoptosis or cause mutations that may lead to cancer.

Regulation of Eukaryotic Cell Cycle

Regulation of the cell cycle involves processes crucial to the survival of a cell, including the detection and repair of genetic damage as well as the prevention of uncontrolled cell division. The molecular events that control the cell cycle are ordered and directional; that is, each process occurs in a sequential fashion and it is impossible to "reverse" the cycle.

Role of Cyclins and CDKs

Nobel Laureate Paul Nurse

Two key classes of regulatory molecules, cyclins and cyclin-dependent kinases (CDKs), determine a cell's progress through the cell cycle. Leland H. Hartwell, R. Timothy Hunt, and Paul M. Nurse won the 2001 Nobel Prize in Physiology or Medicine for their discovery of these central molecules. Many of the genes encoding cyclins and CDKs are conserved among all eukaryotes, but in general more complex organisms have more elaborate cell cycle control systems that incorporate more individual components. Many of the relevant genes were first identified by studying yeast, especially *Saccharomyces cerevisiae*; genetic nomenclature in yeast dubs many of these genes *cdc* (for "cell division cycle") followed by an identifying number, e.g. *cdc25* or *cdc20*.

Cyclins form the regulatory subunits and CDKs the catalytic subunits of an activated heterodimer; cyclins have no catalytic activity and CDKs are inactive in the absence of a partner cyclin. When activated by a bound cyclin, CDKs perform a common biochemical reaction called phosphorylation that activates or inactivates target proteins to orchestrate coordinated entry into the next phase of the cell cycle. Different cyclin-CDK combinations determine the downstream proteins targeted. CDKs are constitutively expressed in cells whereas cyclins are synthesised at specific stages of the cell cycle, in response to various molecular signals.

General Mechanism of Cyclin-CDK Interaction

Upon receiving a pro-mitotic extracellular signal, G_1 cyclin-CDK complexes become active to prepare the cell for S phase, promoting the expression of transcription factors that in turn promote the expression of S cyclins and of enzymes required for DNA replication. The G_1 cyclin-CDK complexes also promote the degradation of molecules that function as S phase inhibitors by targeting them for ubiquitination. Once a protein has been ubiquitinated, it is targeted for proteolytic degradation by the proteasome. However, results from a recent study of E2F transcriptional dynamics at the single-cell level argue that the role of G1 cyclin-CDK activities, in particular cyclin D-CDK4/6, is to tune the timing rather than the commitment of cell cycle entry.

Active S cyclin-CDK complexes phosphorylate proteins that make up the pre-replication complexes assembled during G_1 phase on DNA replication origins. The phosphorylation serves two purposes: to activate each already-assembled pre-replication complex, and to prevent new complexes from forming. This ensures that every portion of the cell's genome will be replicated once and only once. The reason for prevention of gaps in replication is fairly clear, because daughter cells that are missing all or part of crucial genes will die. However, for reasons related to gene copy number effects, possession of extra copies of certain genes is also deleterious to the daughter cells.

Mitotic cyclin-CDK complexes, which are synthesized but inactivated during S and G_2 phases, promote the initiation of mitosis by stimulating downstream proteins involved in chromosome condensation and mitotic spindle assembly. A critical complex activated during this process is a ubiquitin ligase known as the anaphase-promoting complex (APC), which promotes degradation of structural proteins associated with the chromosomal kinetochore. APC also targets the mitotic cyclins for degradation, ensuring that telophase and cytokinesis can proceed.

Specific Action of Cyclin-CDK Complexes

Cyclin D is the first cyclin produced in the cell cycle, in response to extracellular signals (e.g. growth factors). Cyclin D binds to existing CDK4, forming the active cyclin D-CDK4 complex. Cyclin D-CDK4 complex in turn phosphorylates the retinoblastoma susceptibility protein (Rb). The hyperphosphorylated Rb dissociates from the E2F/DP1/Rb complex (which was bound to the E2F responsive genes, effectively "blocking" them from transcription), activating E2F. Activation of E2F results in transcription of various genes like cyclin E, cyclin A, DNA polymerase, thymidine kinase, etc. Cyclin E thus produced binds to CDK2, forming the cyclin E-CDK2 complex, which pushes the cell from G_1 to S phase (G_1/S, which initiates the G_2/M transition). Cyclin B-cdk1 complex activation causes breakdown of nuclear envelope and initiation of prophase, and subsequently, its deactivation causes the cell to exit mitosis.

A quantitative study of E2F transcriptional dynamics at the single-cell level by using engineered fluorescent reporter cells provided a quantitative framework for understanding the control logic of cell cycle entry, challenging the canonical textbook model. Genes that regulate the amplitude of

E2F accumulation, such as Myc, determine the commitment into cell cycle and S phase entry. G1 cyclin-CDK activities are not the driver of cell cycle entry. Instead, they primarily tune the timing of E2F increase, thereby modulating the pace of cell cycle progression.

Inhibitors

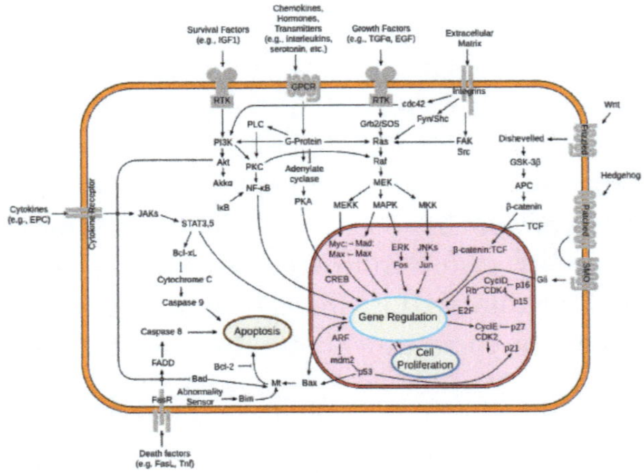

Overview of signal transduction pathways involved in apoptosis, also known as "programmed cell death".

Two families of genes, the *cip/kip* (*CDK interacting protein/Kinase inhibitory protein*) family and the INK4a/ARF (*In*hibitor of *K*inase 4/*A*lternative *R*eading *F*rame) family, prevent the progression of the cell cycle. Because these genes are instrumental in prevention of tumor formation, they are known as tumor suppressors.

The cip/kip family includes the genes p21, p27 and p57. They halt cell cycle in G_1 phase, by binding to, and inactivating, cyclin-CDK complexes. p21 is activated by p53 (which, in turn, is triggered by DNA damage e.g. due to radiation). p27 is activated by Transforming Growth Factor of β (TGF β), a growth inhibitor.

The INK4a/ARF family includes $p16^{INK4a}$, which binds to CDK4 and arrests the cell cycle in G_1 phase, and $p14^{ARF}$ which prevents p53 degradation.

Synthetic inhibitors of Cdc25 could also be useful for the arrest of cell cycle and therefore be useful as antineoplastic and anticancer agents.

Transcriptional Regulatory Network

Current evidence suggests that a semi-autonomous transcriptional network acts in concert with the CDK-cyclin machinery to regulate the cell cycle. Several gene expression studies in *Saccharomyces cerevisiae* have identified 800-1200 genes that change expression over the course of the cell cycle. They are transcribed at high levels at specific points in the cell cycle, and remain at lower levels throughout the rest of the cycle. While the set of identified genes differs between studies due to the computational methods and criteria used to identify them, each study indicates that a large portion of yeast genes are temporally regulated.

Many periodically expressed genes are driven by transcription factors that are also periodically expressed. One screen of single-gene knockouts identified 48 transcription factors (about 20%

of all non-essential transcription factors) that show cell cycle progression defects. Genome-wide studies using high throughput technologies have identified the transcription factors that bind to the promoters of yeast genes, and correlating these findings with temporal expression patterns have allowed the identification of transcription factors that drive phase-specific gene expression. The expression profiles of these transcription factors are driven by the transcription factors that peak in the prior phase, and computational models have shown that a CDK-autonomous network of these transcription factors is sufficient to produce steady-state oscillations in gene expression).

Experimental evidence also suggests that gene expression can oscillate with the period seen in dividing wild-type cells independently of the CDK machinery. Orlando *et al.* used microarrays to measure the expression of a set of 1,271 genes that they identified as periodic in both wild type cells and cells lacking all S-phase and mitotic cyclins (*clb1,2,3,4,5,6*). Of the 1,271 genes assayed, 882 continued to be expressed in the cyclin-deficient cells at the same time as in the wild type cells, despite the fact that the cyclin-deficient cells arrest at the border between G_1 and S phase. However, 833 of the genes assayed changed behavior between the wild type and mutant cells, indicating that these genes are likely directly or indirectly regulated by the CDK-cyclin machinery. Some genes that continued to be expressed on time in the mutant cells were also expressed at different levels in the mutant and wild type cells. These findings suggest that while the transcriptional network may oscillate independently of the CDK-cyclin oscillator, they are coupled in a manner that requires both to ensure the proper timing of cell cycle events. Other work indicates that phosphorylation, a post-translational modification, of cell cycle transcription factors by Cdk1 may alter the localization or activity of the transcription factors in order to tightly control timing of target genes.

While oscillatory transcription plays a key role in the progression of the yeast cell cycle, the CDK-cyclin machinery operates independently in the early embryonic cell cycle. Before the midblastula transition, zygotic transcription does not occur and all needed proteins, such as the B-type cyclins, are translated from maternally loaded mRNA.

DNA Replication and DNA Replication Origin Activity

Analyses of synchronized cultures of *Saccharomyces cerevisiae* under conditions that prevent DNA replication initiation without delaying cell cycle progression showed that origin licensing decreases the expression of genes with origins near their 3' ends, revealing that downstream origins can regulate the expression of upstream genes. This confirms previous predictions from mathematical modeling of a global causal coordination between DNA replication origin activity and mRNA expression, and shows that mathematical modeling of DNA microarray data can be used to correctly predict previously unknown biological modes of regulation.

Checkpoints

Cell cycle checkpoints are used by the cell to monitor and regulate the progress of the cell cycle. Checkpoints prevent cell cycle progression at specific points, allowing verification of necessary phase processes and repair of DNA damage. The cell cannot proceed to the next phase until checkpoint requirements have been met. Checkpoints typically consist of a network of regulatory proteins that monitor and dictate the progression of the cell through the different stages of the cell cycle.

There are several checkpoints to ensure that damaged or incomplete DNA is not passed on to daughter cells. Three main checkpoints exist: the G_1/S checkpoint, the G_2/M checkpoint and the metaphase (mitotic) checkpoint.

G_1/S transition is a rate-limiting step in the cell cycle and is also known as restriction point. This is where the cell checks whether it has enough raw materials to fully replicate its DNA (nucleotide bases, DNA synthase, chromatin, etc.). An unhealthy or malnourished cell will get stuck at this checkpoint.

The G_2/M checkpoint is where the cell ensures that it has enough cytoplasm and phospholipids for two daughter cells. But sometimes more importantly, it checks to see if it is the right time to replicate. There are some situations where many cells need to all replicate simultaneously (for example, a growing embryo should have a symmetric cell distribution until it reaches the mid-blastula transition). This is done by controlling the G_2/M checkpoint.

The metaphase checkpoint is a fairly minor checkpoint, in that once a cell is in metaphase, it has committed to undergoing mitosis. However that's not to say it isn't important. In this checkpoint, the cell checks to ensure that the spindle has formed and that all of the chromosomes are aligned at the spindle equator before anaphase begins.

While these are the three "main" checkpoints, not all cells have to pass through each of these checkpoints in this order to replicate. Many types of cancer are caused by mutations that allow the cells to speed through the various checkpoints or even skip them altogether. Going from S to M to S phase almost consecutively. Because these cells have lost their checkpoints, any DNA mutations that may have occurred are disregarded and passed on to the daughter cells. This is one reason why cancer cells have a tendency to exponentially accrue mutations. Aside from cancer cells, many fully differentiated cell types no longer replicate so they leave the cell cycle and stay in G_0 until their death. Thus removing the need for cellular checkpoints. An alternative model of the cell cycle response to DNA damage has also been proposed, known as the postreplication checkpoint.

Checkpoint regulation plays an important role in an organism's development. In sexual reproduction, when egg fertilization occurs, when the sperm binds to the egg, it releases signalling factors that notify the egg that it has been fertilized. Among other things, this induces the now fertilized oocyte to return from its previously dormant, G_0, state back into the cell cycle and on to mitotic replication and division.

p53 plays an important role in triggering the control mechanisms at both G_1/S and G_2/M checkpoints. In addition to p53, checkpoint regulators are being heavily researched for their roles in cancer growth and proliferation.

Role in Tumor Formation

A disregulation of the cell cycle components may lead to tumor formation. As mentioned above, when some genes like the cell cycle inhibitors, RB, p53 etc. mutate, they may cause the cell to multiply uncontrollably, forming a tumor. Although the duration of cell cycle in tumor cells is equal to or longer than that of normal cell cycle, the proportion of cells that are in active cell division (versus quiescent cells in G_0 phase) in tumors is much higher than that in normal tissue. Thus there is a net increase in cell number as the number of cells that die by apoptosis or senescence remains the same.

The cells which are actively undergoing cell cycle are targeted in cancer therapy as the DNA is relatively exposed during cell division and hence susceptible to damage by drugs or radiation. This fact is made use of in cancer treatment; by a process known as debulking, a significant mass of the tumor is removed which pushes a significant number of the remaining tumor cells from G_0 to G_1 phase (due to increased availability of nutrients, oxygen, growth factors etc.). Radiation or chemotherapy following the debulking procedure kills these cells which have newly entered the cell cycle.

The fastest cycling mammalian cells in culture, crypt cells in the intestinal epithelium, have a cycle time as short as 9 to 10 hours. Stem cells in resting mouse skin may have a cycle time of more than 200 hours. Most of this difference is due to the varying length of G_1, the most variable phase of the cycle. M and S do not vary much.

In general, cells are most radiosensitive in late M and G_2 phases and most resistant in late S phase.

For cells with a longer cell cycle time and a significantly long G_1 phase, there is a second peak of resistance late in G_1.

The pattern of resistance and sensitivity correlates with the level of sulfhydryl compounds in the cell. Sulfhydryls are natural substances that protect cells from radiation damage and tend to be at their highest levels in S and at their lowest near mitosis.

Fluorescence Imaging of the Cell Cycle

Fluorescent proteins visualize the cell cycle progression. IFP2.0-hGem(1/110) fluorescence is shown in green and highlights the S/G_2/M phases. smURFP-hCdtI(30/120) fluorescence is shown in red and highlights the G_0/G_1 phases.

Pioneering work by Atsushi Miyawaki and coworkers developed the fluorescent ubiquitination-based cell cycle indicator (FUCCI), which enables fluorescence imaging of the cell cycle. Originally, a green fluorescent protein, mAG, was fused to hGem(1/110) and an orange fluorescent protein (mKO_2) was fused to hCdt1(30/120). Note, these fusions are fragments that contain a nuclear localization signal and ubiquitination sites for degradation, but are not functional proteins. The green fluorescent protein is made during the S, G_2, or M phase and degraded during the G_0 or G_1 phase, while the orange fluorescent protein is made during the G_0 or G_1 phase and destroyed during the S, G_2, or M phase. A far-red and near-infrared FUCCI was developed using a cyanobacteria-derived fluorescent protein (smURFP) and a bacteriophytochrome-derived fluorescent protein (movie found at this link).

Mitosis

In cell biology, Mitosis is a part of the cell cycle in which chromosomes in a cell nucleus are separated into two identical sets of chromosomes, and each set ends up in its own nucleus. In general, mitosis (division of the nucleus) is preceded by the S stage of interphase (during which the DNA is replicated) and is often accompanied or followed by cytokinesis, which divides the cytoplasm, organelles and cell membrane into two new cells containing roughly equal shares of these cellular components. Mitosis and cytokinesis together define the mitotic (M) phase of an animal cell cycle—the division of the mother cell into two daughter cells, genetically identical to each other and to their parent cell.

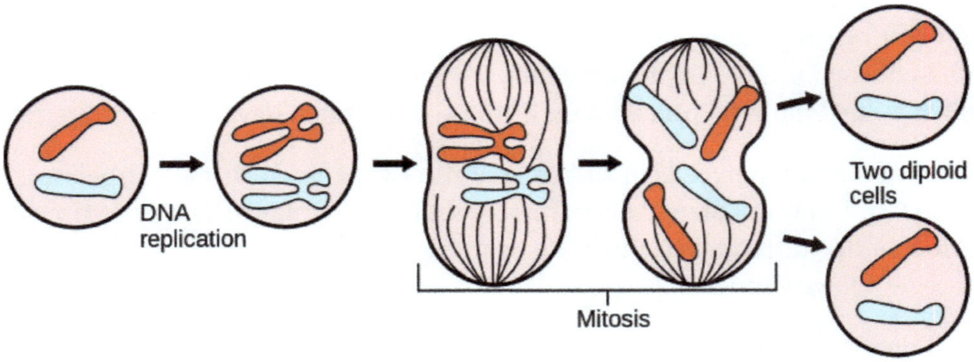

Mitosis divides the chromosomes in a cell nucleus.

The process of mitosis is divided into stages corresponding to the completion of one set of activities and the start of the next. These stages are prophase, prometaphase, metaphase, anaphase, and telophase. During mitosis, the chromosomes, which have already duplicated, condense and attach to spindle fibers that pull one copy of each chromosome to opposite sides of the cell. The result is two genetically identical daughter nuclei. The rest of the cell may then continue to divide by cytokinesis to produce two daughter cells. Producing three or more daughter cells instead of normal two is a mitotic error called tripolar mitosis or multipolar mitosis (direct cell triplication / multiplication). Other errors during mitosis can induce apoptosis (programmed cell death) or cause mutations. Certain types of cancer can arise from such mutations.

Mitosis occurs only in eukaryotic cells and the process varies in different organisms. For example, animals undergo an "open" mitosis, where the nuclear envelope breaks down before the chromosomes separate, while fungi undergo a "closed" mitosis, where chromosomes divide within an intact cell nucleus. Furthermore, most animal cells undergo a shape change, known as mitotic cell rounding, to adopt a near spherical morphology at the start of mitosis. Prokaryotic cells, which lack a nucleus, divide by a different process called binary fission.

Discovery

German zoologist Otto Bütschli might have claimed the discovery of the process presently known as "mitosis", a term coined by Walther Flemming in 1882.

Mitosis was discovered in frog, rabbit, and cat cornea cells in 1873 and described for the first time by the Polish histologist Wacław Mayzel in 1875. The term is derived from the Greek word μίτος *mitos* "warp thread".

Overview of Mitosis

The primary result of mitosis and cytokinesis is the transfer of a parent cell's genome into two daughter cells. The genome is composed of a number of chromosomes—complexes of tightly coiled DNA that contain genetic information vital for proper cell function. Because each resultant daughter cell should be genetically identical to the parent cell, the parent cell must make a copy of each chromosome before mitosis. This occurs during the S phase of interphase. Chromosome duplication results in two identical *sister chromatids* bound together by cohesin proteins at the *centromere*.

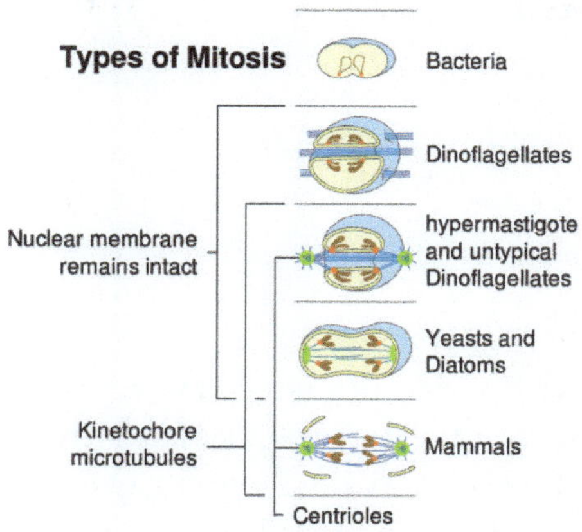

Types of Mitosis

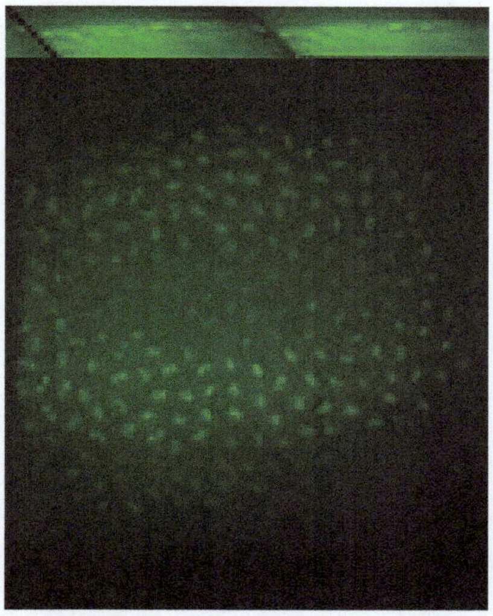

Time-lapse video of mitosis in a *Drosophila melanogaster* embryo.

When mitosis begins, the chromosomes condense and become visible. In some eukaryotes, for example animals, the nuclear envelope, which segregates the DNA from the cytoplasm, disintegrates into small vesicles. The nucleolus, which makes ribosomes in the cell, also disappears. Microtubules project from opposite ends of the cell, attach to the centromeres, and align the chromosomes centrally within the cell. The microtubules then contract to pull the sister chromatids of each chromosome apart. Sister chromatids at this point are called *daughter chromosomes*. As the cell elongates, corresponding daughter chromosomes are pulled toward opposite ends of the cell and condense maximally in late anaphase. A new nuclear envelope forms around the separated daughter chromosomes, which decondense to form interphase nuclei.

During mitotic progression, typically after anaphase onset, the cell may undergo cytokinesis. In animal cells, a cell membrane pinches inward between the two developing nuclei to produce two new cells. In plant cells, a cell plate forms between the two nuclei. Cytokinesis does not always occur; coenocytic (a type of multinucleate condition) cells undergo mitosis without cytokinesis.

Phases of Cell Cycle and Mitosis

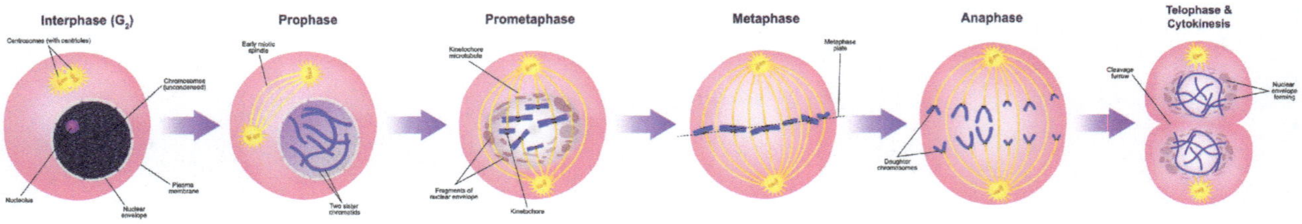

Interphase

The mitotic phase is a relatively short period of the cell cycle. It alternates with the much longer *interphase*, where the cell prepares itself for the process of cell division. Interphase is divided into three phases: G_1 (first gap), S (synthesis), and G_2 (second gap). During all three phases, the cell grows by producing proteins and cytoplasmic organelles. However, chromosomes are replicated only during the S phase. Thus, a cell grows (G_1), continues to grow as it duplicates its chromosomes (S), grows more and prepares for mitosis (G_2), and finally divides (M) before restarting the cycle. All these phases in the cell cycle are highly regulated by cyclins, cyclin-dependent kinases, and other cell cycle proteins. The phases follow one another in strict order and there are "checkpoints" that give the cell cues to proceed from one phase to another. Cells may also temporarily or permanently leave the cell cycle and enter G_0 phase to stop dividing. This can occur when cells become overcrowded (density-dependent inhibition) or when they differentiate to carry out specific functions for the organism, as is the case for human heart muscle cells and neurons. Some G_0 cells have the ability to re-enter the cell cycle.

Preprophase (Plant Cells)

In plant cells only, prophase is preceded by a pre-prophase stage. In highly vacuolated plant cells, the nucleus has to migrate into the center of the cell before mitosis can begin. This is achieved through the formation of a phragmosome, a transverse sheet of cytoplasm that bisects the cell along the future plane of cell division. In addition to phragmosome formation, preprophase is characterized by the formation of a ring of microtubules and actin filaments (called preprophase band) underneath the plasma membrane around the equatorial plane of the future mitotic spindle. This band marks the position where the cell will eventually divide. The cells of higher plants (such as the flowering plants) lack centrioles; instead, microtubules form a spindle on the surface of the nucleus and are then organized into a spindle by the chromosomes themselves, after the nuclear envelope breaks down. The preprophase band disappears during nuclear envelope breakdown and spindle formation in prometaphase.

Prophase

During prophase, which occurs after G_2 interphase, the cell prepares to divide by tightly condensing its chromosomes and initiating mitotic spindle formation, this process is called chromosome condensation. During interphase, the genetic material in the nucleus consists of loosely packed chromatin. At the onset of prophase, chromatin fibers condense into discrete chromosomes that are typically visible at high magnification through a light microscope. In this stage, chromosomes are long, thin and thread-like. Each chromosome has two chromatids. The two chromatids are joined at a place called centromere.

Gene transcription ceases during prophase and does not resume until late anaphase to early G1 phase. The nucleolus also disappears during early prophase.

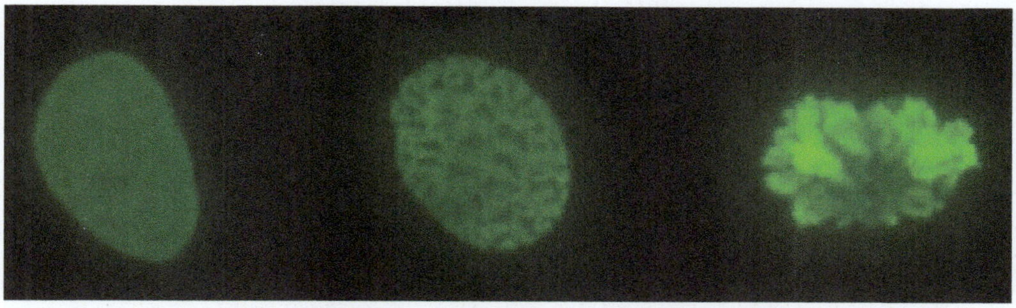

Condensing chromosomes. Interphase nucleus (left), condensing chromosomes (middle) and condensed chromosomes (right).

Close to the nucleus of animal cells are structures called centrosomes, consisting of a pair of centrioles surrounded by a loose collection of proteins. The centrosome is the coordinating center for the cell's microtubules. A cell inherits a single centrosome at cell division, which is duplicated by the cell before a new round of mitosis begins, giving a pair of centrosomes. The two centrosomes polymerize tubulin to help form a microtubule spindle apparatus. Motor proteins then push the centrosomes along these microtubules to opposite sides of the cell. Although centrosomes help organize microtubule assembly, they are not essential for the formation of the spindle apparatus, since they are absent from plants, and are not absolutely required for animal cell mitosis.

Prometaphase

At the beginning of prometaphase in animal cells, phosphorylation of nuclear lamins causes the nuclear envelope to disintegrate into small membrane vesicles. As this happens, microtubules invade the nuclear space. This is called *open mitosis*, and it occurs in some multicellular organisms. Fungi and some protists, such as algae or trichomonads, undergo a variation called *closed mitosis* where the spindle forms inside the nucleus, or the microtubules penetrate the intact nuclear envelope.

In late prometaphase, *kinetochore microtubules* begin to search for and attach to chromosomal kinetochores. A *kinetochore* is a proteinaceous microtubule-binding structure that forms on the chromosomal centromere during late prophase. A number of *polar microtubules* find and interact with corresponding polar microtubules from the opposite centrosome to form the mitotic spindle. Although the kinetochore structure and function are not fully understood, it is known that it contains some form of molecular motor. When a microtubule connects with the kinetochore, the motor activates, using energy from ATP to "crawl" up the tube toward the originating centrosome. This motor activity, coupled with polymerisation and depolymerisation of microtubules, provides the pulling force necessary to later separate the chromosome's two chromatids.

Metaphase

After the microtubules have located and attached to the kinetochores in prometaphase, the two centrosomes begin pulling the chromosomes towards opposite ends of the cell. The resulting tension causes the chromosomes to align along the *metaphase plate* or *equatorial plane*, an imaginary line that is centrally located between the two centrosomes (at approximately the midline of the cell). To ensure equitable distribution of chromosomes at the end of mitosis, the *metaphase checkpoint* guarantees that kinetochores are properly attached to the mitotic spindle and that the chromosomes are aligned along the metaphase plate. If the cell successfully passes through the metaphase checkpoint, it proceeds to anaphase.

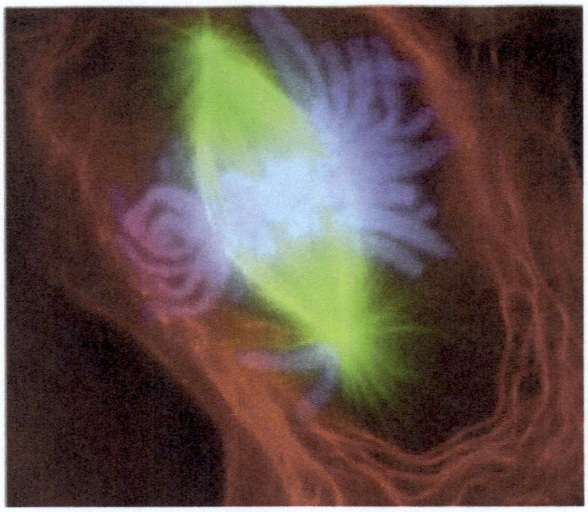

A cell in late metaphase. All chromosomes (blue) but one have arrived at the metaphase plate.

Anaphase

During *anaphase A*, the cohesins that bind sister chromatids together are cleaved, forming two identical daughter chromosomes. Shortening of the kinetochore microtubules pulls the newly formed daughter chromosomes to opposite ends of the cell. During *anaphase B*, polar microtubules push against each other, causing the cell to elongate. In late anaphase, chromosomes also reach their overall maximal condensation level, to help chromosome segregation and the re-formation of the nucleus. In most animal cells, anaphase A precedes anaphase B, but some vertebrate egg cells demonstrate the opposite order of events.

Telophase

Telophase is a reversal of prophase and prometaphase events. At telophase, the polar microtubules continue to lengthen, elongating the cell even more. If the nuclear envelope has broken down, a new nuclear envelope forms using the membrane vesicles of the parent cell's old nuclear envelope. The new envelope forms around each set of separated daughter chromosomes (though the membrane does not enclose the centrosomes) and the nucleolus reappears. Both sets of chromosomes, now surrounded by new nuclear membrane, begin to "relax" or decondense. Mitosis is complete. Each daughter nucleus has an identical set of chromosomes. Cell division may or may not occur at this time depending on the organism.

Cytokinesis

Cytokinesis is not a phase of mitosis but rather a separate process, necessary for completing cell division. In animal cells, a cleavage furrow (pinch) containing a contractile ring develops where the metaphase plate used to be, pinching off the separated nuclei. In both animal and plant cells, cell division is also driven by vesicles derived from the Golgi apparatus, which move along microtubules to the middle of the cell. In plants, this structure coalesces into a cell plate at the center of the phragmoplast and develops into a cell wall, separating the two nuclei. The phragmoplast is a microtubule structure typical for higher plants, whereas some green algae use a phycoplast microtubule array during cytokinesis. Each daughter cell has a complete copy of the genome of its parent cell. The end of cytokinesis marks the end of the M-phase.

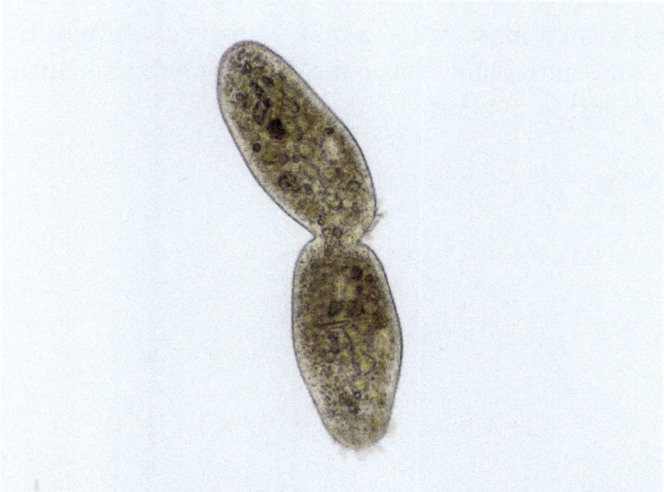

Cilliate undergoing cytokinesis, with the cleavage furrow being clearly visible

There are many cells where mitosis and cytokinesis occur separately, forming single cells with multiple nuclei. The most notable occurrence of this is among the fungi, slime molds, and coenocytic algae, but the phenomenon is found in various other organisms. Even in animals, cytokinesis and mitosis may occur independently, for instance during certain stages of fruit fly embryonic development.

Significance

Mitosis is important for the maintenance of the chromosomal set; each cell formed receives chromosomes that are alike in composition and equal in number to the chromosomes of the parent cell.

Mitosis occurs in the following circumstances:

Development and growth

> The number of cells within an organism increases by mitosis. This is the basis of the development of a multicellular body from a single cell, i.e., zygote and also the basis of the growth of a multicellular body.

Cell replacement

> In some parts of body, e.g. skin and digestive tract, cells are constantly sloughed off and replaced by new ones. New cells are formed by mitosis and so are exact copies of the cells being replaced. In like manner, red blood cells have short lifespan (only about 4 months) and new RBCs are formed by mitosis.

Regeneration

Some organisms can regenerate body parts. The production of new cells in such instances is achieved by mitosis. For example, starfish regenerate lost arms through mitosis.

Asexual reproduction

Some organisms produce genetically similar offspring through asexual reproduction. For example, the hydra reproduces asexually by budding. The cells at the surface of hydra undergo mitosis and form a mass called a bud. Mitosis continues in the cells of the bud and this grows into a new individual. The same division happens during asexual reproduction or vegetative propagation in plants.

Cell Shape Changes During Mitosis

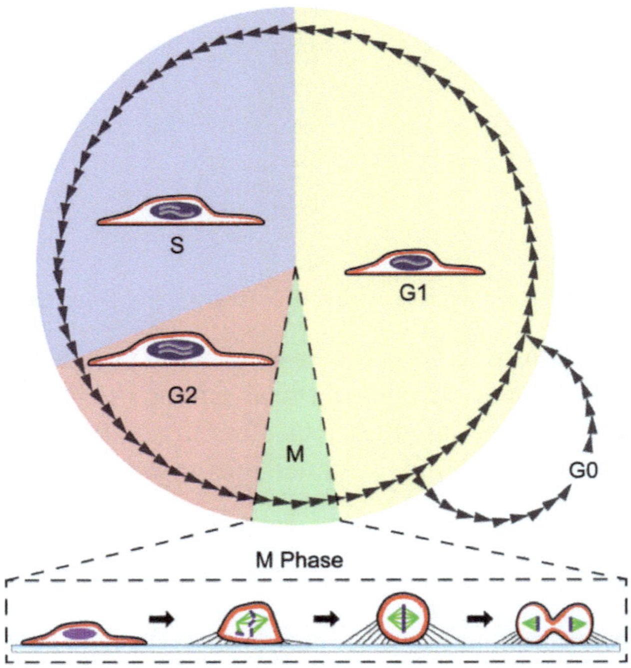

Cell shape changes through mitosis for a typical animal cell cultured on a flat surface. The cell undergoes mitotic cell rounding during spindle assembly and then divides via cytokinesis. The actomyosin cortex is depicted in red, DNA/chromosomes purple, microtubules green, and membrane and retraction fibers in black. Rounding also occurs in live tissue, as described in the text.

In animal tissue, most cells round up to a near-spherical shape during mitosis. In epithelia and epidermis, an efficient rounding process is correlated with proper mitotic spindle alignment and subsequent correct positioning of daughter cells. Moreover, researchers have found that if rounding is heavily suppressed it may result in spindle defects, primarily pole splitting and failure to efficiently capture chromosomes. Therefore, mitotic cell rounding is thought to play a protective role in ensuring accurate mitosis.

Rounding forces are driven by reorganization of F-actin and myosin (actomyosin) into a contractile homogeneous cell cortex that 1) rigidifies the cell periphery and 2) facilitates generation of intracellular hydrostatic pressure (up to 10 fold higher than interphase). The generation of intracellular pressure is particularly critical under confinement, such as would be important in a tissue scenario, where outward forces must be produced to round up against surrounding cells and/or the extracellular matrix. Generation of pressure is dependent on formin-mediated F-actin nucleation and Rho kinase (ROCK)-mediated myosin II contraction, both of which are governed upstream by signaling pathways RhoA and ECT2 through the activity of Cdk1. Due to its importance in mitosis, the molecular components and dynamics of the mitotic actomyosin cortex is an area of active research.

Errors and Variations of Mitosis

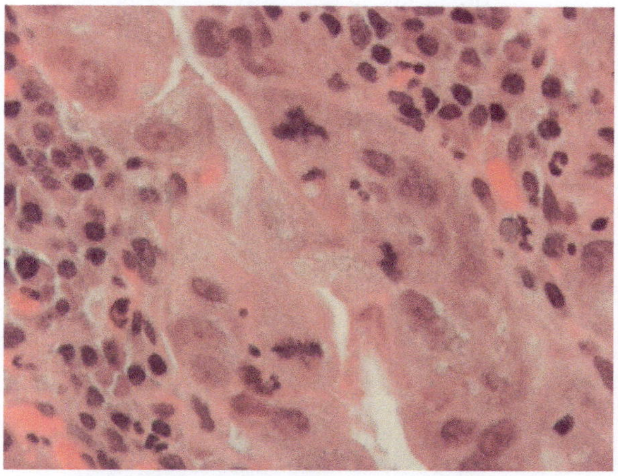

An abnormal (tripolar) mitosis (12 o'clock position) in a precancerous lesion of the stomach (H&E stain)

Errors can occur during mitosis, especially during early embryonic development in humans. Mitotic errors can create aneuploid cells that have too few or too many of one or more chromosomes, a condition associated with cancer. Early human embryos, cancer cells, infected or intoxicated cells can also suffer from pathological division into three or more daughter cells (tripolar or multipolar mitosis), resulting in severe errors in their chromosomal complements.

In *nondisjunction*, sister chromatids fail to separate during anaphase. One daughter cell receives both sister chromatids from the nondisjoining chromosome and the other cell receives none. As a result, the former cell gets three copies of the chromosome, a condition known as *trisomy*, and the latter will have only one copy, a condition known as *monosomy*. On occasion, when cells experience nondisjunction, they fail to complete cytokinesis and retain both nuclei in one cell, resulting in binucleated cells.

Anaphase lag occurs when the movement of one chromatid is impeded during anaphase. This may be caused by a failure of the mitotic spindle to properly attach to the chromosome. The lagging chromatid is excluded from both nuclei and is lost. Therefore, one of the daughter cells will be monosomic for that chromosome.

Endoreduplication (or endoreplication) occurs when chromosomes duplicate but the cell does not subsequently divide. This results in polyploid cells or, if the chromosomes duplicates repeatedly, polytene chromosomes. Endoreduplication is found in many species and appears to be a normal part of development. Endomitosis is a variant of endoreduplication in which cells replicate their chromosomes during S phase and enter, but prematurely terminate, mitosis. Instead of being divided into two new daughter nuclei, the replicated chromosomes are retained within the original nucleus. The cells then re-enter G_1 and S phase and replicate their chromosomes again. This may occur multiple times, increasing the chromosome number with each round of replication and endomitosis. Platelet-producing megakaryocytes go through endomitosis during cell differentiation.

Timeline in Pictures

Mitotic cells can be visualized microscopically by staining them with fluorescent antibodies and dyes.

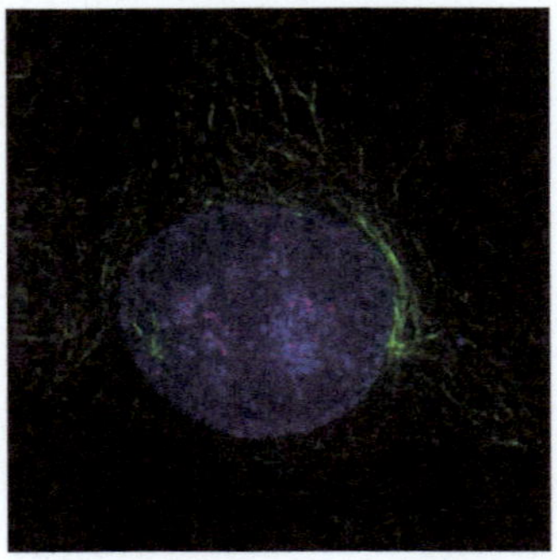

Early prophase: Nonkinetochore microtubules, shown as green strands, have established a matrix around the disassembling nucleus, with the condensing chromosomes in blue. The red nodules are the centromeres.

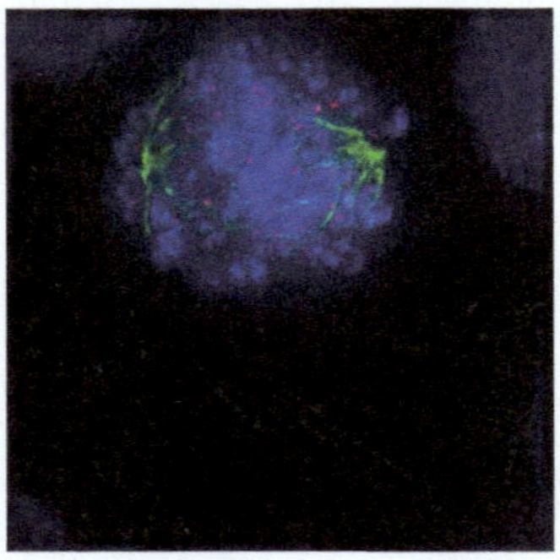

Early prometaphase: The nuclear envelope has just disassembled, allowing the microtubules to quickly interact with the kinetochores, which assemble on the centromeres of the condensing chromosomes.

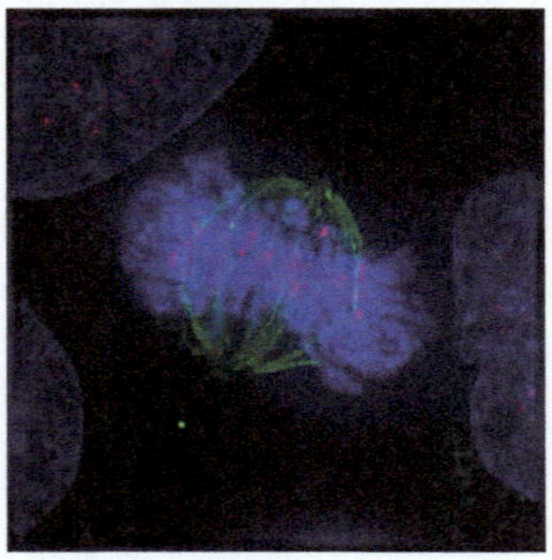

Metaphase: The centrosomes have moved to the poles of the cell and have established the mitotic spindle. The chromosomes have congressed at the metaphase plate.

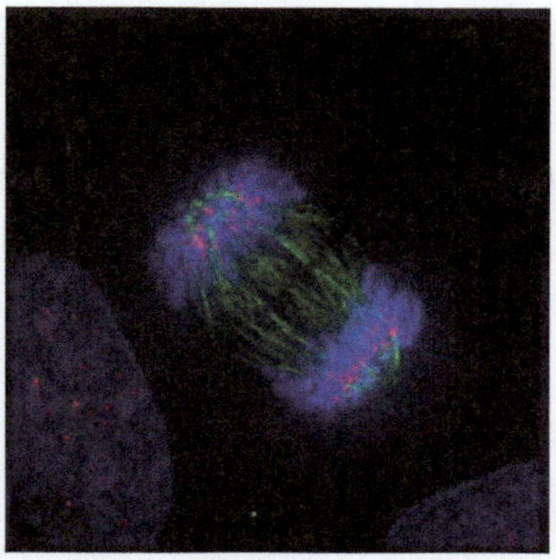

Anaphase: Kinetochore microtubules pull the two sets of chromosomes apart, and lengthening nonkinetochore microtubules push the halves of the dividing cell further apart, while chromosomes are condensed maximally.

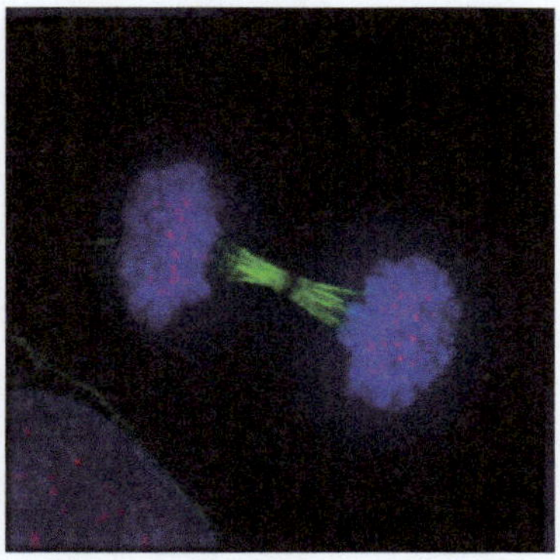

Telophase: Reversal of prophase and prometaphase events and thus completing the cell cycle.

Cell-cycle Analysis

Cell-cycle analysis is a method in cell biology that employs flow cytometry to distinguish cells in different phases of the cell cycle. Before analysis, the cells are permeabilised and treated with a fluorescent dye that stains DNA quantitatively, usually propidium iodide (PI). The fluorescence intensity of the stained cells at certain wavelengths will therefore correlate with the amount of DNA they contain. As the DNA content of cells duplicates during the S phase of the cell cycle, the relative amount of cells in the G_0 phase and G_1 phase (before S phase), in the S phase, and in the G_2 phase and M phase (after S phase) can be determined, as the fluorescence of cells in the G_2/M phase will be twice as high as that of cells in the G_0/G_1 phase.

Cell-cycle anomalies can be symptoms for various kinds of cell damage, for example DNA damage, which cause the cell to interrupt the cell cycle at certain checkpoints to prevent transformation into a cancer cell (carcinogenesis). Other possible reasons for anomalies include lack of nutrients, for example after serum deprivation.

Cell cycle analysis was first described in 1969 at Los Alamos Scientific Laboratory by a group from the University of California, using the Feulgen staining technique. The first protocol for cell-cycle analysis using propidium iodide staining was presented in 1975 by Awtar Krishan from Harvard Medical School and is still widely cited today.

Experimental Procedure

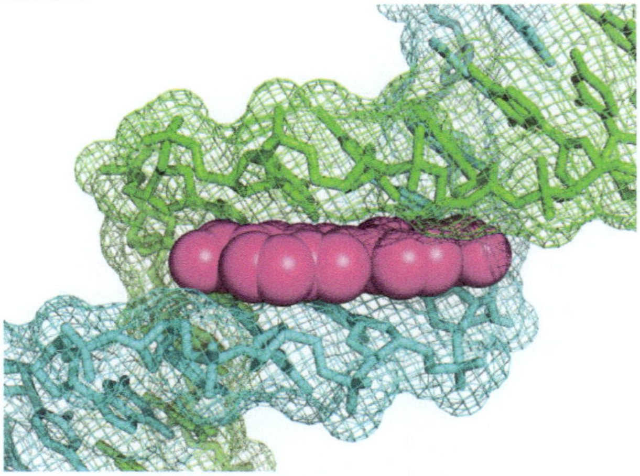

DAPI (magenta) bound to the minor groove of DNA (green and blue). From PDB: 1D30.

The first step in preparing cells for cell-cycle analysis is permeabilisation of the cells' plasma membranes. This is usually done by incubating them in a buffer solution containing a detergent such as Triton X-100 or NP-40, or by fixating them in ethanol. Most fluorescent DNA dyes are not membrane permeable, that is, unable to pass through an intact cell membrane. Permeabilisation is therefore crucial for the success of the next step, the staining of the cells.

Prior to or during the staining step, the cells will usually be treated with RNase A to remove RNAs from the cells. This is important because dyes that stain DNA will also stain RNA, thus creating artefacts that would distort the results.

Aside from propidium iodide, quantifiable dyes that are frequently used include (but are not limited to) DRAQ5, 7-Aminoactinomycin D, DAPI and Hoechst 33342.

When the cells pass through the flow cytometer's laser, a fluorescence pulse is generated that correlates with the amount of dye associated with the DNA and thus with the total amount of DNA in the cell.

Doublet Discrimination

Because cells and especially fixated cells tend to stick together, cell aggregates have to be excluded from analysis through a process called *doublet discrimination*. This is important because a doublet of two cells in the G_0/G_1 phase has the same total amount of DNA and thus the same fluorescence intensity as a single cell in the G_2/M phase. G_0/G_1 doublets would therefore create false positive results for G_2/M cells.

Related Methods

Nicoletti Assay

The Nicoletti assay, named after its inventor, the Italian physician Ildo Nicoletti, is a modified form of cell cycle analysis. It is used to detect and measure apoptosis, a form of programmed cell death, by analysing cells with a DNA content less than 2n ("sub-G_1 cells"). Such cells are usually the result of apoptotic DNA fragmentation: during apoptosis, the DNA is degraded by cellular endonucleases. Therefore, nuclei of apoptotic cells contain less DNA than nuclei of healthy G_0/G_1 cells, resulting in a sub-G_0/G_1 peak in the fluorescence histogram that can be used to determine the relative amount of apoptotic cells in a sample.

This method was developed and first described in 1991 by Nicoletti and co-workers at Perugia University School of Medicine. An optimised protocol developed by two of the authors of the original publication was published in 2006.

Nicoletti assay with healthy (left) and apoptotic (middle and right) cells

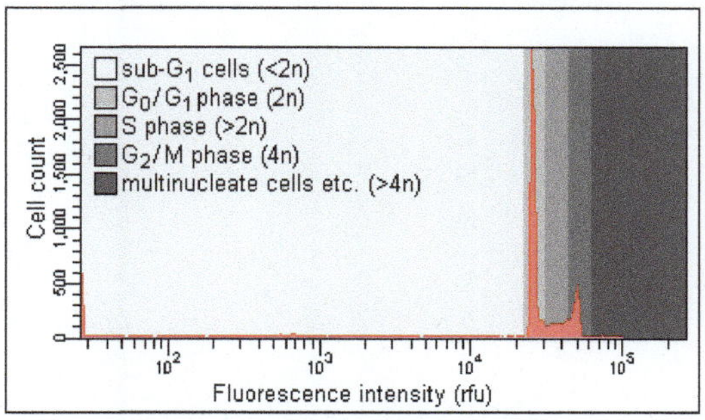

Healthy cells. Note the absence of a sub-G_0/G_1 peak.

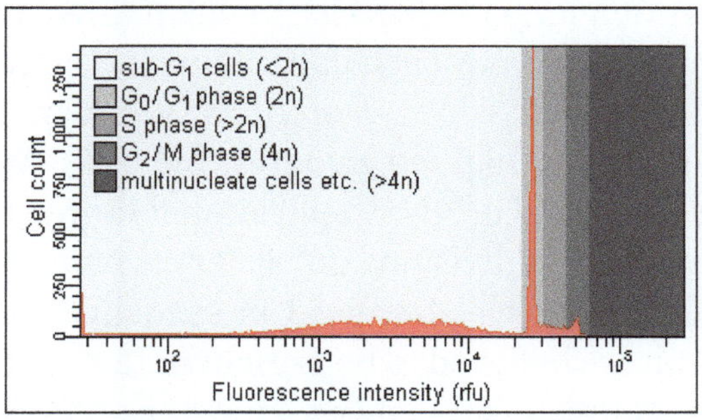

Apoptotic cells one day after apoptosis induction. Note the presence of a sub-G_0/G_1 peak.

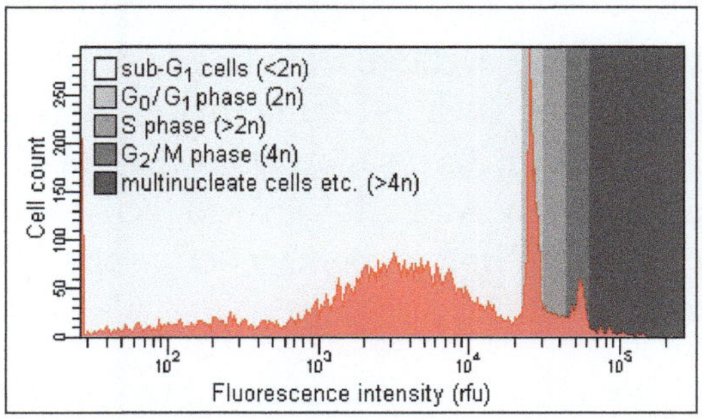

Apoptotic cells several days after apoptosis induction. Note the relative increase of the sub-G_0/G_1 peak.

Cell Division

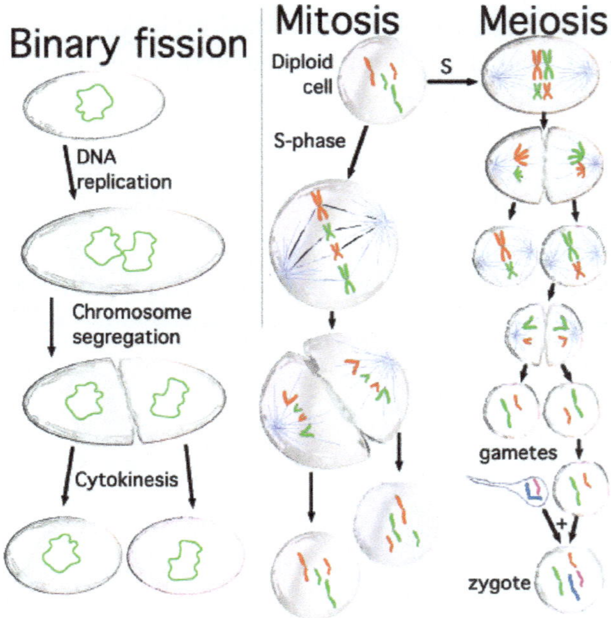

Three types of cell division

Cell division is the process by which a parent cell divides into two or more daughter cells. Cell division usually occurs as part of a larger cell cycle. In eukaryotes, there are two distinct types of cell division: a vegetative division, whereby each daughter cell is genetically identical to the parent cell (mitosis), and a reproductive cell division, whereby the number of chromosomes in the daughter cells is reduced by half, to produce haploid gametes (meiosis). Meiosis results in four haploid daughter cells by undergoing one round of DNA replication followed by two divisions: homologous chromosomes are separated in the first division, and sister chromatids are separated in the second division. Both of these cell division cycles are used in sexually reproducing organisms at some point in their life cycle, and both are believed to be present in the last eukaryotic common ancestor. Prokaryotes also undergo a vegetative cell division known as binary fission, where their genetic material is segregated equally into two daughter cells. All cell divisions, regardless of organism, are preceded by a single round of DNA replication.

For simple unicellular organisms[Note 1] such as the amoeba, one cell division is equivalent to reproduction – an entire new organism is created. On a larger scale, mitotic cell division can create progeny from multicellular organisms, such as plants that grow from cuttings. Cell division also enables sexually reproducing organisms to develop from the one-celled zygote, which itself was produced by cell division from gametes. And after growth, cell division allows for continual construction and repair of the organism. A human being's body experiences about 10 quadrillion cell divisions in a lifetime.

The primary concern of cell division is the maintenance of the original cell's genome. Before division can occur, the genomic information that is stored in chromosomes must be replicated, and the duplicated genome must be separated cleanly between cells. A great deal of cellular infrastructure is involved in keeping genomic information consistent between "generations".

Variants

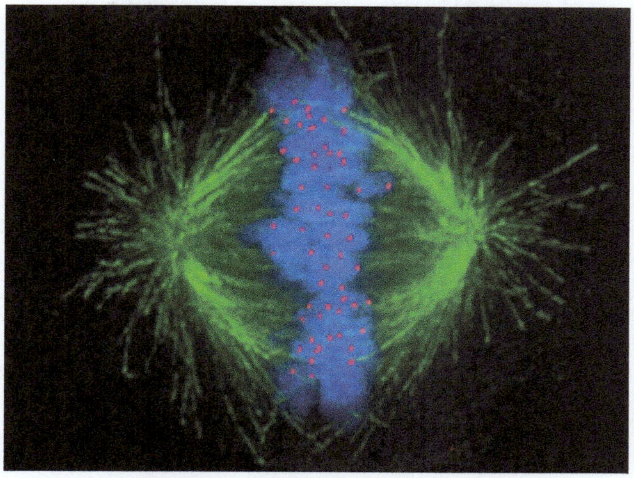

Image of the mitotic spindle in a human cell showing microtubules in green, chromosomes (DNA) in blue, and kinetochores in red.

Cells are classified into two main categories: simple, non-nucleated prokaryotic cells, and complex, nucleated eukaryotic cells. Owing to their structural differences, eukaryotic and prokaryotic cells do not divide in the same way. Also, the pattern of cell division that transforms eukaryotic stem cells into gametes (sperm cells in males or ova – egg cells – in females) is different from that of the somatic cell division in the cells of the body.

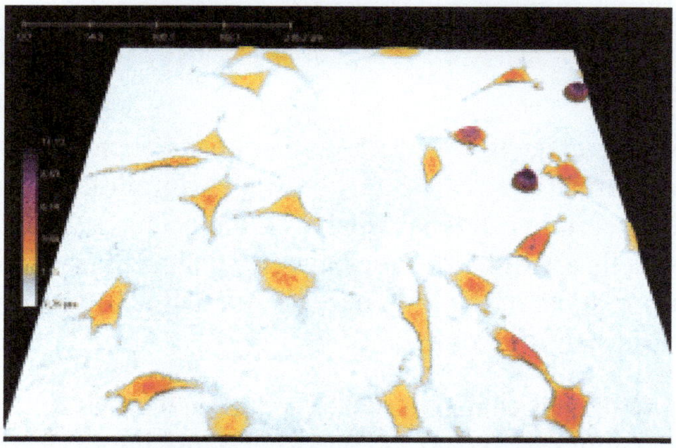

Cell division over 42 hours. The cells were directly imaged in the cell culture vessel, using non-invasive quantitative phase contrast time-lapse microscopy.

Degradation

Multicellular organisms replace worn-out cells through cell division. In some animals, however, cell division eventually halts. In humans this occurs on average, after 52 divisions, known as the Hayflick limit. The cell is then referred to as senescent. Cells stop dividing because the telomeres, protective bits of DNA on the end of a chromosome required for replication, shorten with each copy, eventually being consumed. Cancer cells, on the other hand, are not thought to degrade in this way, if at all. An enzyme called telomerase, present in large quantities in cancerous cells, rebuilds the telomeres, allowing division to continue indefinitely.

Cellular Differentiation

In developmental biology, cellular differentiation is the process where a cell changes from one cell type to another. Most commonly this is a less specialized type becoming a more specialized type, such as during cell growth. Differentiation occurs numerous times during the development of a multicellular organism as it changes from a simple zygote to a complex system of tissues and cell types. Differentiation continues in adulthood as adult stem cells divide and create fully differentiated daughter cells during tissue repair and during normal cell turnover. Some differentiation occurs in response to antigen exposure. Differentiation dramatically changes a cell's size, shape, membrane potential, metabolic activity, and responsiveness to signals. These changes are largely due to highly controlled modifications in gene expression and are the study of epigenetics. With a few exceptions, cellular differentiation almost never involves a change in the DNA sequence itself. Thus, different cells can have very different physical characteristics despite having the same genome.

A cell that can differentiate into all cell types of the adult organism is known as *pluripotent*. Such cells are called embryonic stem cells in animals and meristematic cells in higher plants. A cell that can differentiate into all cell types, including the placental tissue, is known as *totipotent*. In mammals, only the zygote and subsequent blastomeres are totipotent, while in plants many differentiated cells can become totipotent with simple laboratory techniques. In cytopathology, the level of cellular differentiation is used as a measure of cancer progression. "Grade" is a marker of how differentiated a cell in a tumor is.

Mammalian Cell Types

Three basic categories of cells make up the mammalian body: germ cells, somatic cells, and stem cells. Each of the approximately 100 trillion (10^{14}) cells in an adult human has its own copy or copies of the genome except certain cell types, such as red blood cells, that lack nuclei in their fully differentiated state. Most cells are diploid; they have two copies of each chromosome. Such cells, called somatic cells, make up most of the human body, such as skin and muscle cells. Cells differentiate to specialize for different functions.

Germ line cells are any line of cells that give rise to gametes—eggs and sperm—and thus are continuous through the generations. Stem cells, on the other hand, have the ability to divide for indefinite periods and to give rise to specialized cells. They are best described in the context of normal human development.

Development begins when a sperm fertilizes an egg and creates a single cell that has the potential to form an entire organism. In the first hours after fertilization, this cell divides into identical cells. In humans, approximately four days after fertilization and after several cycles of cell division, these cells begin to specialize, forming a hollow sphere of cells, called a blastocyst. The blastocyst has an outer layer of cells, and inside this hollow sphere, there is a cluster of cells called the inner cell mass. The cells of the inner cell mass go on to form virtually all of the tissues of the human body. Although the cells of the inner cell mass can form virtually every type of cell found in the human body, they cannot form an organism. These cells are referred to as pluripotent.

Pluripotent stem cells undergo further specialization into multipotent progenitor cells that then give rise to functional cells. Examples of stem and progenitor cells include:

- *Radial glial cells* (embryonic neural stem cells) that give rise to excitatory neurons in the fetal brain through the process of neurogenesis.
- Hematopoietic stem cells (adult stem cells) from the bone marrow that give rise to red blood cells, white blood cells, and platelets
- *Mesenchymal stem cells* (adult stem cells) from the bone marrow that give rise to stromal cells, fat cells, and types of bone cells
- *Epithelial stem cells* (progenitor cells) that give rise to the various types of skin cells
- *Muscle satellite cells* (progenitor cells) that contribute to differentiated muscle tissue.

A pathway that is guided by the cell adhesion molecules consisting of four amino acids, arginine, glycine, asparagine, and serine, is created as the cellular blastomere differentiates from the single-layered blastula to the three primary layers of germ cells in mammals, namely the ectoderm, mesoderm and endoderm (listed from most distal (exterior) to proximal (interior)). The ectoderm ends up forming the skin and the nervous system, the mesoderm forms the bones and muscular tissue, and the endoderm forms the internal organ tissues.

Dedifferentiation

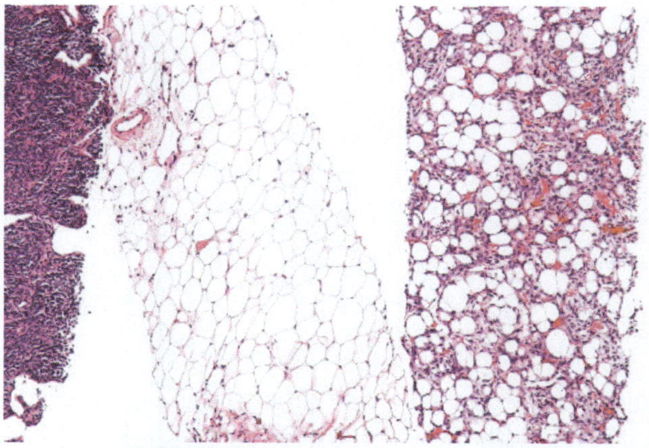

Micrograph of a liposarcoma with some dedifferentiation, that is not identifiable as a liposarcoma, (left edge of image) and a differentiated component (with lipoblasts and increased vascularity (right of image)). Fully differentiated (morphologically benign) adipose tissue (center of the image) has few blood vessels. H&E stain.

Dedifferentiation, or integration is a cellular process often seen in more basal life forms such as worms and amphibians in which a partially or terminally differentiated cell reverts to an earlier developmental stage, usually as part of a regenerative process. Dedifferentiation also occurs in plants. Cells in cell culture can lose properties they originally had, such as protein expression, or change shape. This process is also termed dedifferentiation.

Some believe dedifferentiation is an aberration of the normal development cycle that results in cancer, whereas others believe it to be a natural part of the immune response lost by humans at some point as a result of evolution.

A small molecule dubbed reversine, a purine analog, has been discovered that has proven to induce dedifferentiation in myotubes. These dedifferentiated cells could then redifferentiate into osteoblasts and adipocytes.

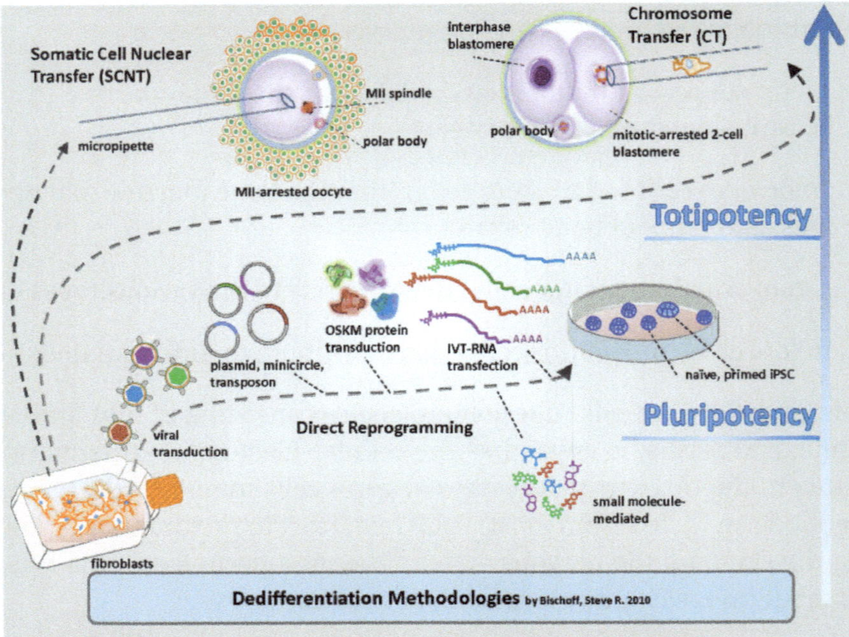

Diagram exposing several methods used to revert adult somatic cells to totipotency or pluripotency.

Mechanisms

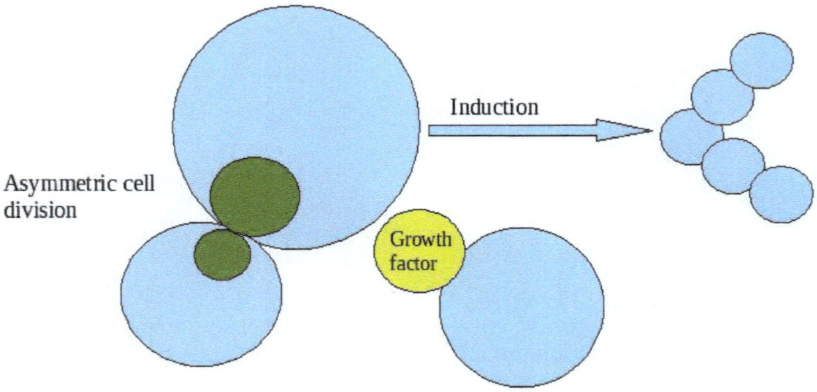

Mechanisms of cellular differentiation.

Each specialized cell type in an organism expresses a subset of all the genes that constitute the genome of that species. Each cell type is defined by its particular pattern of regulated gene expression. Cell differentiation is thus a transition of a cell from one cell type to another and it involves a switch from one pattern of gene expression to another. Cellular differentiation during development can be understood as the result of a gene regulatory network. A regulatory gene and its cis-regulatory modules are nodes in a gene regulatory network; they receive input and create output elsewhere in the network. The systems biology approach to developmental biology emphasizes the importance of investigating how developmental mechanisms interact to produce predictable patterns (morphogenesis). (However, an alternative view has been proposed recently. Based on stochastic gene expression, cellular differentiation is the result of a Darwinian selective process occurring among cells. In this frame, protein and gene networks are the result of cellular processes and not their cause.)

A few evolutionarily conserved types of molecular processes are often involved in the cellular mechanisms that control these switches. The major types of molecular processes that control cellular differentiation involve cell signaling. Many of the signal molecules that convey information from cell to cell during the control of cellular differentiation are called growth factors. Although the details of specific signal transduction pathways vary, these pathways often share the following general steps. A ligand produced by one cell binds to a receptor in the extracellular region of another cell, inducing a conformational change in the receptor. The shape of the cytoplasmic domain of the receptor changes, and the receptor acquires enzymatic activity. The receptor then catalyzes reactions that phosphorylate other proteins, activating them. A cascade of phosphorylation reactions eventually activates a dormant transcription factor or cytoskeletal protein, thus contributing to the differentiation process in the target cell. Cells and tissues can vary in competence, their ability to respond to external signals.

Signal induction refers to cascades of signaling events, during which a cell or tissue signals to another cell or tissue to influence its developmental fate. Yamamoto and Jeffery investigated the role of the lens in eye formation in cave- and surface-dwelling fish, a striking example of induction. Through reciprocal transplants, Yamamoto and Jeffery found that the lens vesicle of surface fish can induce other parts of the eye to develop in cave- and surface-dwelling fish, while the lens vesicle of the cave-dwelling fish cannot.

Other important mechanisms fall under the category of asymmetric cell divisions, divisions that give rise to daughter cells with distinct developmental fates. Asymmetric cell divisions can occur because of asymmetrically expressed maternal cytoplasmic determinants or because of signaling. In the former mechanism, distinct daughter cells are created during cytokinesis because of an uneven distribution of regulatory molecules in the parent cell; the distinct cytoplasm that each daughter cell inherits results in a distinct pattern of differentiation for each daughter cell. A well-studied example of pattern formation by asymmetric divisions is body axis patterning in Drosophila. RNA molecules are an important type of intracellular differentiation control signal. The molecular and genetic basis of asymmetric cell divisions has also been studied in green algae of the genus *Volvox*, a mo del system for studying how unicellular organisms can evolve into multicellular organisms. In *Volvox carteri*, the 16 cells in the anterior hemisphere of a 32-cell embryo divide asymmetrically, each producing one large and one small daughter cell. The size of the cell at the end of all cell divisions determines whether it becomes a specialized germ or somatic cell.

Epigenetic Control Of Cellular Differentiation

Since each cell, regardless of cell type, possesses the same genome, determination of cell type must occur at the level of gene expression. While the regulation of gene expression can occur through cis- and trans-regulatory elements including a gene's promoter and enhancers, the problem arises as to how this expression pattern is maintained over numerous generations of cell division. As it turns out, epigenetic processes play a crucial role in regulating the decision to adopt a stem, progenitor, or mature cell fate. This section will focus primarily on mammalian stem cells.

In systems biology and mathematical modeling of gene regulatory networks, cell-fate determination is predicted to exhibit certain dynamics, such as attractor-convergence (the attractor can be an equilibrium point, limit cycle or strange attractor) or oscillatory.

Importance of Epigenetic Control

The first question that can be asked is the extent and complexity of the role of epigenetic processes in the determination of cell fate. A clear answer to this question can be seen in the 2011 paper by Lister R, *et al.* on aberrant epigenomic programming in human induced pluripotent stem cells. As induced pluripotent stem cells (iPSCs) are thought to mimic embryonic stem cells in their pluripotent properties, few epigenetic differences should exist between them. To test this prediction, the authors conducted whole-genome profiling of DNA methylation patterns in several human embryonic stem cell (ESC), iPSC, and progenitor cell lines.

Female adipose cells, lung fibroblasts, and foreskin fibroblasts were reprogrammed into induced pluripotent state with the OCT4, SOX2, KLF4, and MYC genes. Patterns of DNA methylation in ESCs, iPSCs, somatic cells were compared. Lister R, *et al.* observed significant resemblance in methylation levels between embryonic and induced pluripotent cells. Around 80% of CG dinucleotides in ESCs and iPSCs were methylated, the same was true of only 60% of CG dinucleotides in somatic cells. In addition, somatic cells possessed minimal levels of cytosine methylation in non-CG dinucleotides, while induced pluripotent cells possessed similar levels of methylation as embryonic stem cells, between 0.5 and 1.5%. Thus, consistent with their respective transcriptional activities, DNA methylation patterns, at least on the genomic level, are similar between ESCs and iPSCs.

However, upon examining methylation patterns more closely, the authors discovered 1175 regions of differential CG dinucleotide methylation between at least one ES or iPS cell line. By comparing these regions of differential methylation with regions of cytosine methylation in the original somatic cells, 44-49% of differentially methylated regions reflected methylation patterns of the respective progenitor somatic cells, while 51-56% of these regions were dissimilar to both the progenitor and embryonic cell lines. In vitro-induced differentiation of iPSC lines saw transmission of 88% and 46% of hyper and hypo-methylated differentially methylated regions, respectively.

Two conclusions are readily apparent from this study. First, epigenetic processes are heavily involved in cell fate determination, as seen from the similar levels of cytosine methylation between induced pluripotent and embryonic stem cells, consistent with their respective patterns of transcription. Second, the mechanisms of de-differentiation (and by extension, differentiation) are very complex and cannot be easily duplicated, as seen by the significant number of differentially methylated regions between ES and iPS cell lines. Now that these two points have been established, we can examine some of the epigenetic mechanisms that are thought to regulate cellular differentiation.

Mechanisms of Epigenetic Regulation

Pioneering Factors (Oct4, Sox2, Nanog)

Three transcription factors, OCT4, SOX2, and NANOG – the first two of which are used in iPSC reprogramming – are highly expressed in undifferentiated embryonic stem cells and are necessary for the maintenance of their pluripotency. It is thought that they achieve this through alterations in chromatin structure, such as histone modification and DNA methylation, to restrict or permit the transcription of target genes.

Polycomb Repressive Complex (PRC2)

In the realm of gene silencing, Polycomb repressive complex 2, one of two classes of the Polycomb group (PcG) family of proteins, catalyzes the di- and tri-methylation of histone H3 lysine 27 (H3K27me2/me3). By binding to the H3K27me2/3-tagged nucleosome, PRC1 (also a complex of PcG family proteins) catalyzes the mono-ubiquitinylation of histone H2A at lysine 119 (H2AK119Ub1), blocking RNA polymerase II activity and resulting in transcriptional suppression. PcG knockout ES cells do not differentiate efficiently into the three germ layers, and deletion of the PRC1 and PRC2 genes leads to increased expression of lineage-affiliated genes and unscheduled differentiation. Presumably, PcG complexes are responsible for transcriptionally repressing differentiation and development-promoting genes.

Trithorax Group Proteins (TrxG)

Alternately, upon receiving differentiation signals, PcG proteins are recruited to promoters of pluripotency transcription factors. PcG-deficient ES cells can begin differentiation but cannot maintain the differentiated phenotype. Simultaneously, differentiation and development-promoting genes are activated by Trithorax group (TrxG) chromatin regulators and lose their repression. TrxG proteins are recruited at regions of high transcriptional activity, where they catalyze the trimethylation of histone H3 lysine 4 (H3K4me3) and promote gene activation through histone acetylation. PcG and TrxG complexes engage in direct competition and are thought to be functionally antagonistic, creating at differentiation and development-promoting loci what is termed a "bivalent domain" and rendering these genes sensitive to rapid induction or repression.

DNA Methylation

Regulation of gene expression is further achieved through DNA methylation, in which the DNA methyltransferase-mediated methylation of cytosine residues in CpG dinucleotides maintains heritable repression by controlling DNA accessibility. The majority of CpG sites in embryonic stem cells are unmethylated and appear to be associated with H3K4me3-carrying nucleosomes. Upon differentiation, a small number of genes, including OCT4 and NANOG, are methylated and their promoters repressed to prevent their further expression. Consistently, DNA methylation-deficient embryonic stem cells rapidly enter apoptosis upon in vitro differentiation.

Nucleosome Positioning

While the DNA sequence of most cells of an organism is the same, the binding patterns of transcription factors and the corresponding gene expression patterns are different. To a large extent, differences in transcription factor binding are determined by the chromatin accessibility of their binding sites through histone modification and/or pioneer factors. In particular, it is important to know whether a nucleosome is covering a given genomic binding site or not. Recent studies have elucidated the role of nucleosome positioning during stem cell development.

Role of Signaling in Epigenetic Control

A final question to ask concerns the role of cell signaling in influencing the epigenetic processes governing differentiation. Such a role should exist, as it would be reasonable to think that extrinsic signaling can lead to epigenetic remodeling, just as it can lead to changes in gene expression through the activation or repression of different transcription factors. Interestingly, little direct data is available concerning the specific signals that influence the epigenome, and the majority of current knowledge consist of speculations on plausible candidate regulators of epigenetic remodeling. We will first discuss several major candidates thought to be involved in the induction and maintenance of both embryonic stem cells and their differentiated progeny, and then turn to one example of specific signaling pathways in which more direct evidence exists for its role in epigenetic change.

The first major candidate is Wnt signaling pathway. The Wnt pathway is involved in all stages of differentiation, and the ligand Wnt3a can substitute for the overexpression of c-Myc in the generation of induced pluripotent stem cells. On the other hand, disruption of ß-catenin, a component of the Wnt signaling pathway, leads to decreased proliferation of neural progenitors.

Growth factors comprise the second major set of candidates of epigenetic regulators of cellular differentiation. These morphogens are crucial for development, and include bone morphogenetic proteins, transforming growth factors (TGFs), and fibroblast growth factors (FGFs). TGFs and FGFs have been shown to sustain expression of OCT4, SOX2, and NANOG by downstream signaling to Smad proteins. Depletion of growth factors promotes the differentiation of ESCs, while genes with bivalent chromatin can become either more restrictive or permissive in their transcription.

Several other signaling pathways are also considered to be primary candidates. Cytokine leukemia inhibitory factors are associated with the maintenance of mouse ESCs in an undifferentiated state. This is achieved through its activation of the Jak-STAT3 pathway, which has been shown to be necessary and sufficient towards maintaining mouse ESC pluripotency. Retinoic acid can induce differentiation of human and mouse ESCs, and Notch signaling is involved in the proliferation and self-renewal of stem cells. Finally, Sonic hedgehog, in addition to its role as a morphogen, promotes embryonic stem cell differentiation and the self-renewal of somatic stem cells.

The problem, of course, is that the candidacy of these signaling pathways was inferred primarily on the basis of their role in development and cellular differentiation. While epigenetic regulation is necessary for driving cellular differentiation, they are certainly not sufficient for this process. Direct modulation of gene expression through modification of transcription factors plays a key role that must be distinguished from heritable epigenetic changes that can persist even in the absence of the original environmental signals. Only a few examples of signaling pathways leading to epigenetic changes that alter cell fate currently exist, and we will focus on one of them.

Expression of Shh (Sonic hedgehog) upregulates the production of BMI1, a component of the PcG complex that recognizes H3K27me3. This occurs in a Gli-dependent manner, as Gli1 and Gli2 are downstream effectors of the Hedgehog signaling pathway. In culture, Bmi1 mediates the Hedgehog pathway's ability to promote human mammary stem cell self-renewal. In both humans and mice, researchers showed Bmi1 to be highly expressed in proliferating immature cerebellar granule cell precursors. When Bmi1 was knocked out in mice, impaired cerebellar development resulted, leading to significant reductions in postnatal brain mass along with abnormalities in motor control and behavior. A separate study showed a significant decrease in neural stem cell proliferation along with increased astrocyte proliferation in Bmi null mice.

In summary, the role of signaling in the epigenetic control of cell fate in mammals is largely unknown, but distinct examples exist that indicate the likely existence of further such mechanisms.

Effect of Matrix Elasticity on Differentiation

In order to fulfill the purpose of regenerating a variety of tissues, adult stems are known to migrate from their niches, adhere to new extracellular matrices (ECM) and differentiate. The ductility of these microenvironments are unique to different tissue types. The ECM surrounding brain, muscle and bone tissues range from soft to stiff. The transduction of the stem cells into these cells types is not directed solely by chemokine cues and cell to cell signaling. The elasticity of the microenvironment can also affect the differentiation of mesenchymal stem cells (MSCs which originate in bone marrow.) When MSCs are placed on substrates of the same stiffness as brain, muscle and bone ECM, the MSCs take on properties of those respective cell types. Matrix sensing requires the cell to pull against the matrix at focal adhesions, which triggers a cellular mechano-transducer to generate a signal to be informed what force is needed to deform the matrix. To determine the key players in matrix-elasticity-driven lineage specification in MSCs, different matrix microenvironments were mimicked. From these experiments, it was concluded that focal adhesions of the MSCs were the cellular mechano-transducer sensing the differences of the matrix elasticity. The non-muscle myosin IIa-c isoforms generates the forces in the cell that lead to signaling of early commitment markers. Nonmuslce myosin IIa generates the least force increasing to non-muscle myosin IIc. There are also factors in the cell that inhibit non-muscle myosin II, such as blebbistatin. This makes the cell effectively blind to the surrounding matrix. Researchers have obtained some success in inducing stem cell-like properties in HEK 239 cells by providing a soft matrix without the use of diffusing factors. The stem-cell properties appear to be linked to tension in the cells' actin network. One identified mechanism for matrix-induced differentiation is tension-induced proteins, which remodel chromatin in response to mechanical stretch. The RhoA pathway is also implicated in this process.

DNA Replication

In molecular biology, DNA replication is the biological process of producing two identical replicas of DNA from one original DNA molecule. This process occurs in all living organisms and is the basis for biological inheritance. DNA is made up of a double helix of two complementary strands. During replication, these strands are separated. Each strand of the original DNA molecule then serves as a template for the production of its counterpart, a process referred to as semiconservative replication. Cellular proofreading and error-checking mechanisms ensure near perfect fidelity for DNA replication.

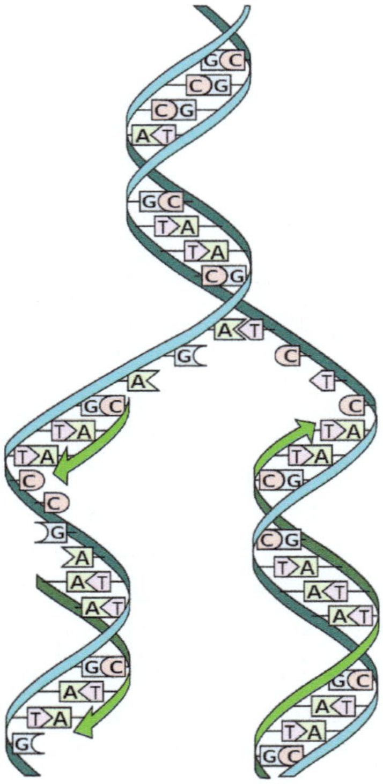

DNA replication. The double helix is unwound and each strand acts as a template (blue) for the next strand. Bases are matched to synthesize the new partner strands (green).

In a cell, DNA replication begins at specific locations, or origins of replication, in the genome. Unwinding of DNA at the origin and synthesis of new strands results in replication forks growing bi-directionally from the origin. A number of proteins are associated with the replication fork to help in the initiation and continuation of DNA synthesis. Most prominently, DNA polymerase synthesizes the new strands by adding nucleotides that complement each (template) strand. DNA replication occurs during the S-stage of interphase.

DNA replication can also be performed *in vitro* (artificially, outside a cell). DNA polymerases isolated from cells and artificial DNA primers can be used to initiate DNA synthesis at known sequences in a template DNA molecule. The polymerase chain reaction (PCR), a common laboratory technique, cyclically applies such artificial synthesis to amplify a specific target DNA fragment from a pool of DNA.

DNA Structures

DNA usually exists as a double-stranded structure, with both strands coiled together to form the characteristic double-helix. Each single strand of DNA is a chain of four types of nucleotides. Nucleotides in DNA contain a deoxyribose sugar, a phosphate, and a nucleobase. The four types of nucleotide correspond to the four nucleobases adenine, cytosine, guanine, and thymine, commonly abbreviated as A,C, G and T. Adenine and guanine are purine bases, while cytosine and thymine are pyrimidines. These nucleotides form phosphodiester bonds, creating the phosphate-deoxyribose backbone of the DNA double helix with the nuclei bases pointing inward (i.e., toward the opposing strand). Nucleotides (bases) are matched between strands through hydrogen bonds to form base pairs. Adenine pairs with thymine (two hydrogen bonds), and guanine pairs with cytosine (stronger: three hydrogen bonds).

DNA strands have a directionality, and the different ends of a single strand are called the "3' (three-prime) end" and the "5' (five-prime) end". By convention, if the base sequence of a single strand of DNA is given, the left end of the sequence is the 5' end, while the right end of the sequence is the 3' end. The strands of the double helix are anti-parallel with one being 5' to 3', and the opposite strand 3' to 5'. These terms refer to the carbon atom in deoxyribose to which the next phosphate in the chain attaches. Directionality has consequences in DNA synthesis, because DNA polymerase can synthesize DNA in only one direction by adding nucleotides to the 3' end of a DNA strand.

The pairing of complementary bases in DNA (through hydrogen bonding) means that the information contained within each strand is redundant. Phosphodiester (intra-strand) bonds are stronger than hydrogen (inter-strand) bonds. This allows the strands to be separated from one another. The nucleotides on a single strand can therefore be used to reconstruct nucleotides on a newly synthesized partner strand.

DNA Polymerase

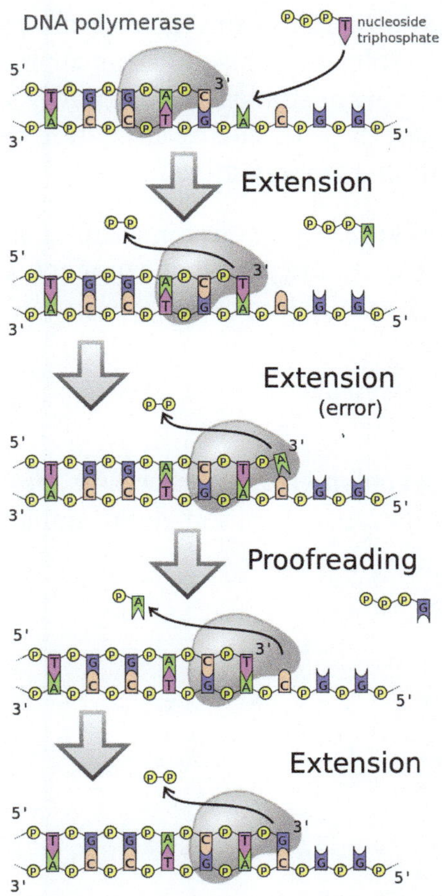

DNA polymerases adds nucleotides to the 3' end of a strand of DNA. If a mismatch is accidentally incorporated, the polymerase is inhibited from further extension. Proofreading removes the mismatched nucleotide and extension continues.

DNA polymerases are a family of enzymes that carry out all forms of DNA replication. DNA polymerases in general cannot initiate synthesis of new strands, but can only extend an existing DNA or RNA strand paired with a template strand. To begin synthesis, a short fragment of RNA, called a primer, must be created and paired with the template DNA strand.

DNA polymerase adds a new strand of DNA by extending the 3' end of an existing nucleotide chain, adding new nucleotides matched to the template strand one at a time via the creation of phosphodiester bonds. The energy for this process of DNA polymerization comes from hydrolysis of the high-energy phosphate (phosphoanhydride) bonds between the three phosphates attached to each unincorporated base. (Free bases with their attached phosphate groups are called nucleotides; in particular, bases with three attached phosphate groups are called nucleoside triphosphates.) When a nucleotide is being added to a growing DNA strand, the formation of a phosphodiester bond between the proximal phosphate of the nucleotide to the growing chain is accompanied by hydrolysis of a high-energy phosphate bond with release of the two distal phosphates as a pyrophosphate. Enzymatic hydrolysis of the resulting pyrophosphate into inorganic phosphate consumes a second high-energy phosphate bond and renders the reaction effectively irreversible.[Note 1]

In general, DNA polymerases are highly accurate, with an intrinsic error rate of less than one mistake for every 10^7 nucleotides added. In addition, some DNA polymerases also have proofreading ability; they can remove nucleotides from the end of a growing strand in order to correct mismatched bases. Finally, post-replication mismatch repair mechanisms monitor the DNA for errors, being capable of distinguishing mismatches in the newly synthesized DNA strand from the original strand sequence. Together, these three discrimination steps enable replication fidelity of less than one mistake for every 10^9 nucleotides added.

The rate of DNA replication in a living cell was first measured as the rate of phage T4 DNA elongation in phage-infected E. coli. During the period of exponential DNA increase at 37 °C, the rate was 749 nucleotides per second. The mutation rate per base pair per replication during phage T4 DNA synthesis is 1.7 per 10^8.

Replication Process

DNA replication, like all biological polymerization processes, proceeds in three enzymatically catalyzed and coordinated steps: initiation, elongation and termination.

Initiation

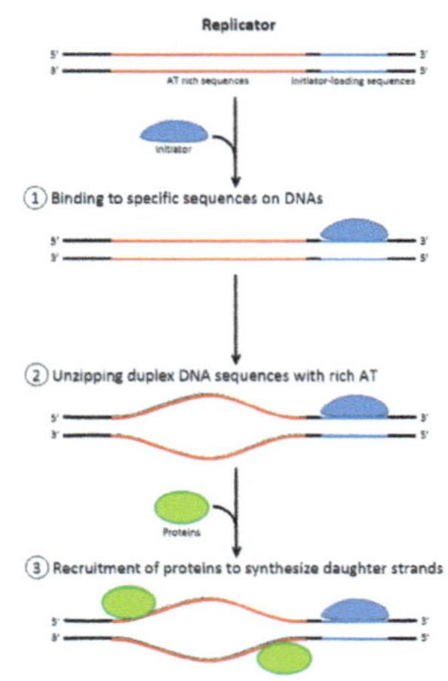

Role of initiators for initiation of DNA replication.

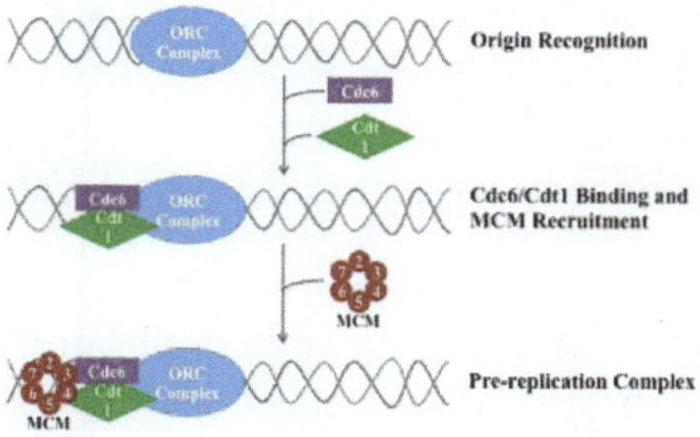

Formation of pre-replication complex.

For a cell to divide, it must first replicate its DNA. This process is initiated at particular points in the DNA, known as “origins”, which are targeted by initiator proteins. In *E. coli* this protein is DnaA; in yeast, this is the origin recognition complex. Sequences used by initiator proteins tend to be “AT-rich” (rich in adenine and thymine bases), because A-T base pairs have two hydrogen bonds (rather than the three formed in a C-G pair) and thus are easier to decouple. Once the origin has been located, these initiators recruit other proteins and form the pre-replication complex, which unzips the double-stranded DNA.

Elongation

DNA polymerase has 5’-3’ activity. All known DNA replication systems require a free 3’ hydroxyl group before synthesis can be initiated (Important note: the DNA template is read in 3’ to 5’ direction whereas a new strand is synthesized in the 5’ to 3’ direction—this is often confused). Four distinct mechanisms for initiation of synthesis are recognized:

1. All cellular life forms and many DNA viruses, phages and plasmids use a primase to synthesize a short RNA primer with a free 3’ OH group which is subsequently elongated by a DNA polymerase.

2. The retroelements (including retroviruses) employ a transfer RNA that primes DNA replication by providing a free 3′ OH that is used for elongation by the reverse transcriptase.

3. In the adenoviruses and the φ29 family of bacteriophages, the 3’ OH group is provided by the side chain of an amino acid of the genome attached protein (the terminal protein) to which nucleotides are added by the DNA polymerase to form a new strand.

4. In the single stranded DNA viruses — a group that includes the circoviruses, the geminiviruses, the parvoviruses and others — and also the many phages and plasmids that use the rolling circle replication (RCR) mechanism, the RCR endonuclease creates a nick in the genome strand (single stranded viruses) or one of the DNA strands (plasmids). The 5′ end of the nicked strand is transferred to a tyrosine residue on the nuclease and the free 3′ OH group is then used by the DNA polymerase to synthesize the new strand.

The first is the best known of these mechanisms and is used by the cellular organisms. In this mechanism, once the two strands are separated, primase adds RNA primers to the template strands. The leading strand receives one RNA primer while the lagging strand receives several. The leading strand is continuously extended from the primer by a DNA polymerase with high processivity, while the lagging strand is extended discontinuously from each primer forming Okazaki fragments. RNase removes the primer RNA fragments, and a low processivity DNA polymerase distinct from the replicative polymerase enters to fill the gaps. When this is complete, a single nick on the leading strand and several nicks on the lagging strand can be found. Ligase works to fill these nicks in, thus completing the newly replicated DNA molecule.

The primase used in this process differs significantly between bacteria and archaea/eukaryotes. Bacteria use a primase belonging to the DnaG protein superfamily which contains a catalytic domain of the TOPRIM fold type. The TOPRIM fold contains an α/β core with four conserved strands in a Rossmann-like topology. This structure is also found in the catalytic domains of topoisomerase Ia, topoisomerase II, the OLD-family nucleases and DNA repair proteins related to the RecR protein.

The primase used by archaea and eukaryotes, in contrast, contains a highly derived version of the RNA recognition motif (RRM). This primase is structurally similar to many viral RNA-dependent RNA polymerases, reverse transcriptases, cyclic nucleotide generating cyclases and DNA polymerases of the A/B/Y families that are involved in DNA replication and repair. In eukaryotic replication, the primase forms a complex with Pol α.

Multiple DNA polymerases take on different roles in the DNA replication process. In *E. coli*, DNA Pol III is the polymerase enzyme primarily responsible for DNA replication. It assembles into a replication complex at the replication fork that exhibits extremely high processivity, remaining intact for the entire replication cycle. In contrast, DNA Pol I is the enzyme responsible for replacing RNA primers with DNA. DNA Pol I has a 5' to 3' exonuclease activity in addition to its polymerase activity, and uses its exonuclease activity to degrade the RNA primers ahead of it as it extends the DNA strand behind it, in a process called nick translation. Pol I is much less processive than Pol III because its primary function in DNA replication is to create many short DNA regions rather than a few very long regions.

In eukaryotes, the low-processivity enzyme, Pol α, helps to initiate replication. The high-processivity extension enzymes are Pol δ and Pol ε.

As DNA synthesis continues, the original DNA strands continue to unwind on each side of the bubble, forming a replication fork with two prongs. In bacteria, which have a single origin of replication on their circular chromosome, this process creates a "theta structure" (resembling the Greek letter theta: θ). In contrast, eukaryotes have longer linear chromosomes and initiate replication at multiple origins within these.>

Replication Fork

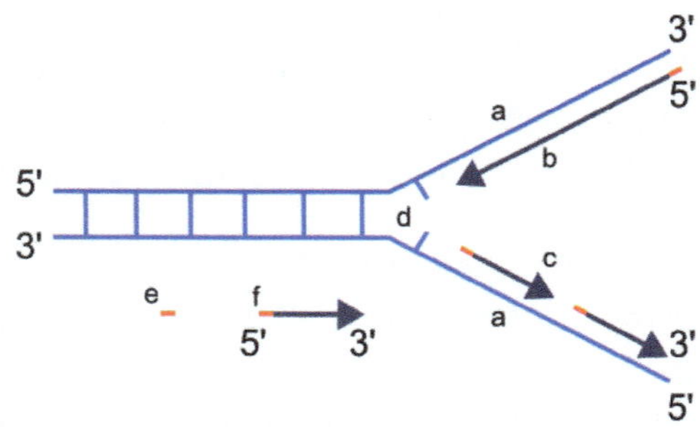

Scheme of the replication fork.
a: template, b: leading strand, c: lagging strand, d: replication fork, e: primer, f: Okazaki fragments

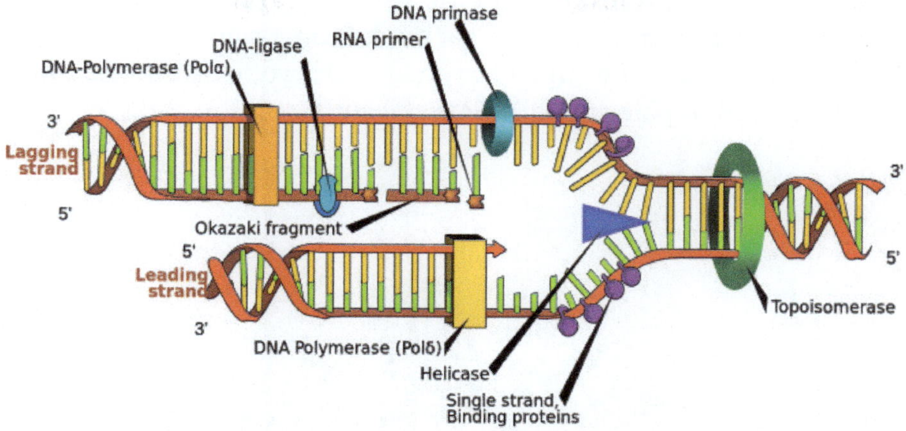

Many enzymes are involved in the DNA replication fork.

The replication fork is a structure that forms within the nucleus during DNA replication. It is created by helicases, which break the hydrogen bonds holding the two DNA strands together. The resulting structure has two branching "prongs", each one made up of a single strand of DNA. These two strands serve as the template for the leading and lagging strands, which will be created as DNA polymerase matches complementary nucleotides to the templates; the templates may be properly referred to as the leading strand template and the lagging strand template.

DNA is always synthesized in the 5' to 3' direction. Since the leading and lagging strand templates are oriented in opposite directions at the replication fork, a major issue is how to achieve synthesis of nascent (new) lagging strand DNA, whose direction of synthesis is opposite to the direction of the growing replication fork.

Leading Strand

The leading strand is the strand of nascent DNA which is being synthesized in the same direction as the growing replication fork. A polymerase "reads" the leading strand *template* and adds complementary nucleotides to the nascent leading strand on a continuous basis.

The polymerase involved in leading strand synthesis is DNA polymerase III (DNA Pol III) in prokaryotes. In eukaryotes, leading strand synthesis is thought to be conducted by Pol ε, however this view has been recently challenged, suggesting a role for Pol δ.

Lagging Strand

The lagging strand is the strand of nascent DNA whose direction of synthesis is opposite to the direction of the growing replication fork. Because of its orientation, replication of the lagging strand is more complicated as compared to that of the leading strand.

The lagging strand is synthesized in short, separated segments. On the lagging strand *template*, a primase "reads" the template DNA and initiates synthesis of a short complementary RNA primer. A DNA polymerase extends the primed segments, forming Okazaki fragments. The RNA primers are then removed and replaced with DNA, and the fragments of DNA are joined together by DNA ligase.

DNA polymerase III (in prokaryotes) or Pol δ (in eukaryotes) is responsible for extension of the primers added during replication of the lagging strand. Primer removal is performed by DNA polymerase I (in prokaryotes) and Pol δ (in eukaryotes). Eukaryotic primase is intrinsic to Pol α. In eukaryotes, pol ε helps with repair during DNA replication.

Dynamics at the Replication Fork

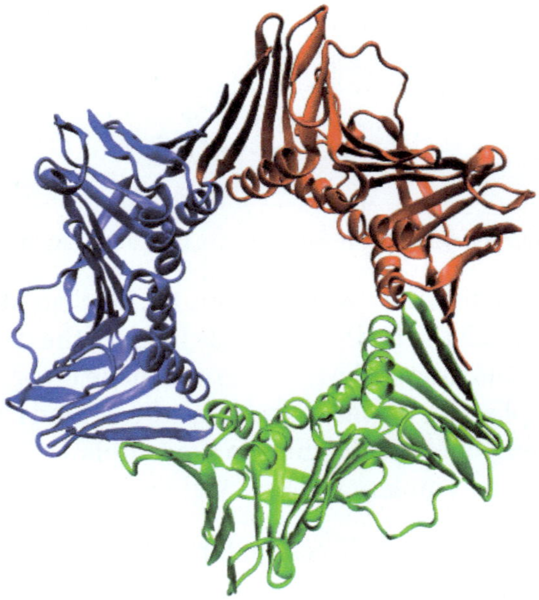

The assembled human DNA clamp, a trimer of the protein PCNA.

As helicase unwinds DNA at the replication fork, the DNA ahead is forced to rotate. This process results in a build-up of twists in the DNA ahead. This build-up forms a torsional resistance that would eventually halt the progress of the replication fork. Topoisomerases are enzymes that temporarily break the strands of DNA, relieving the tension caused by unwinding the two strands of the DNA helix; topoisomerases (including DNA gyrase) achieve this by adding negative supercoils to the DNA helix.

Bare single-stranded DNA tends to fold back on itself forming secondary structures; these structures can interfere with the movement of DNA polymerase. To prevent this, single-strand binding proteins bind to the DNA until a second strand is synthesized, preventing secondary structure formation.

Clamp proteins form a sliding clamp around DNA, helping the DNA polymerase maintain contact with its template, thereby assisting with processivity. The inner face of the clamp enables DNA to be threaded through it. Once the polymerase reaches the end of the template or detects double-stranded DNA, the sliding clamp undergoes a conformational change that releases the DNA polymerase. Clamp-loading proteins are used to initially load the clamp, recognizing the junction between template and RNA primers.

DNA Replication Proteins

At the replication fork, many replication enzymes assemble on the DNA into a complex molecular machine called the replisome. The following is a list of major DNA replication enzymes that participate in the replisome:

Enzyme	Function in DNA replication
DNA Helicase	Also known as helix destabilizing enzyme. Unwinds the DNA double helix at the Replication Fork.
DNA Polymerase	Builds a new duplex DNA strand by adding nucleotides in the 5' to 3' direction. Also performs proof-reading and error correction. There exist many different types of DNA Polymerase, each of which perform different functions in different types of cells.
DNA clamp	A protein which prevents elongating DNA polymerases from dissociating from the DNA parent strand.
Single-Strand Binding (SSB) Proteins	Bind to ssDNA and prevent the DNA double helix from re-annealing after DNA helicase unwinds it, thus maintaining the strand separation, and facilitating the synthesis of the nascent strand.
Topoisomerase	Relaxes the DNA from its super-coiled nature.
DNA Gyrase	Relieves strain of unwinding by DNA helicase; this is a specific type of topoisomerase
DNA Ligase	Re-anneals the semi-conservative strands and joins Okazaki Fragments of the lagging strand.
Primase	Provides a starting point of RNA (or DNA) for DNA polymerase to begin synthesis of the new DNA strand.
Telomerase	Lengthens telomeric DNA by adding repetitive nucleotide sequences to the ends of eukaryotic chromosomes. This allows germ cells and stem cells to avoid the Hayflick limit on cell division.

Replication Machinery

Replication machineries consist of factors involved in DNA replication and appearing on template ssDNAs. Replication machineries include primosotors are replication enzymes; DNA polymerase, DNA helicases, DNA clamps and DNA topoisomerases, and replication proteins; e.g. single-stranded DNA binding proteins (SSB). In the replication machineries these components coordinate. In most of the bacteria, all of the factors involved in DNA replication are located on replication forks and the complexes stay on the forks during DNA replication. These replication machineries are called replisomes or DNA replicase systems. These terms are generic terms for proteins located on replication forks. In eukaryotic and some bacterial cells the replisomes are not formed.

Since replication machineries do not move relatively to template DNAs such as factories, they are called a **replication factory**. In an alternative figure, DNA factories are similar to projectors and

DNAs are like as cinematic films passing constantly into the projectors. In the replication factory model, after both DNA helicases for leading stands and lagging strands are loaded on the template DNAs, the helicases run along the DNAs into each other. The helicases remain associated for the remainder of replication process. Peter Meister et al. observed directly replication sites in budding yeast by monitoring green fluorescent protein(GFP)-tagged DNA polymerases α. They detected DNA replication of pairs of the tagged loci spaced apart symmetrically from a replication origin and found that the distance between the pairs decreased markedly by time. This finding suggests that the mechanism of DNA replication goes with DNA factories. That is, couples of replication factories are loaded on replication origins and the factories associated with each other. Also, template DNAs move into the factories, which bring extrusion of the template ssDNAs and nascent DNAs. Meister's finding is the first direct evidence of replication factory model. Subsequent research has shown that DNA helicases form dimers in many eukaryotic cells and bacterial replication machineries stay in single intranuclear location during DNA synthesis.

The replication factories perform disentanglement of sister chromatids. The disentanglement is essential for distributing the chromatids into daughter cells after DNA replication. Because sister chromatids after DNA replication hold each other by **Cohesin** rings, there is the only chance for the disentanglement in DNA replication. Fixing of replication machineries as replication factories can improve the success rate of DNA replication. If replication forks move freely in chromosomes, catenation of nuclei is aggravated and impedes mitotic segregation.

Termination

Eukaryotes initiate DNA replication at multiple points in the chromosome, so replication forks meet and terminate at many points in the chromosome; these are not known to be regulated in any particular way. Because eukaryotes have linear chromosomes, DNA replication is unable to reach the very end of the chromosomes, but ends at the telomere region of repetitive DNA close to the ends. This shortens the telomere of the daughter DNA strand. Shortening of the telomeres is a normal process in somatic cells. As a result, cells can only divide a certain number of times before the DNA loss prevents further division. (This is known as the Hayflick limit.) Within the germ cell line, which passes DNA to the next generation, telomerase extends the repetitive sequences of the telomere region to prevent degradation. Telomerase can become mistakenly active in somatic cells, sometimes leading to cancer formation. Increased telomerase activity is one of the hallmarks of cancer.

Termination requires that the progress of the DNA replication fork must stop or be blocked. Termination at a specific locus, when it occurs, involves the interaction between two components: (1) a termination site sequence in the DNA, and (2) a protein which binds to this sequence to physically stop DNA replication. In various bacterial species, this is named the DNA replication terminus site-binding protein, or Ter protein.

Because bacteria have circular chromosomes, termination of replication occurs when the two replication forks meet each other on the opposite end of the parental chromosome. *E. coli* regulates this process through the use of termination sequences that, when bound by the Tus protein, enable only one direction of replication fork to pass through. As a result, the replication forks are constrained to always meet within the termination region of the chromosome.

Regulation

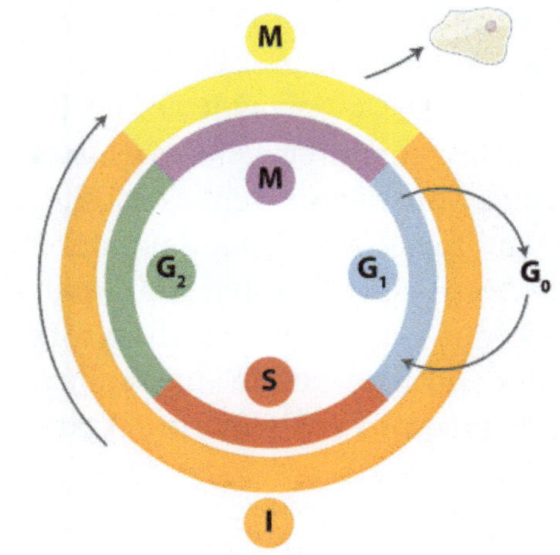

The cell cycle of eukaryotic cells.

Eukaryotes

Within eukaryotes, DNA replication is controlled within the context of the cell cycle. As the cell grows and divides, it progresses through stages in the cell cycle; DNA replication takes place during the S phase (synthesis phase). The progress of the eukaryotic cell through the cycle is controlled by cell cycle checkpoints. Progression through checkpoints is controlled through complex interactions between various proteins, including cyclins and cyclin-dependent kinases. Unlike bacteria, eukaryotic DNA replicates in the confines of the nucleus.

The G1/S checkpoint (or restriction checkpoint) regulates whether eukaryotic cells enter the process of DNA replication and subsequent division. Cells that do not proceed through this checkpoint remain in the G0 stage and do not replicate their DNA.

Replication of chloroplast and mitochondrial genomes occurs independently of the cell cycle, through the process of D-loop replication.

Replication Focus

In vertebrate cells, replication sites concentrate into positions called **replication foci**. Replication sites can be detected by immunostaining daughter strands and replication enzymes and monitoring GFP-tagged replication factors. By these methods it is found that replication foci of varying size and positions appear in S phase of cell division and their number per nucleus is far smaller than the number of genomic replication forks.

P. Heun et al.(2001) tracked GFP-tagged replication foci in budding yeast cells and revealed that replication origins move constantly in G1 and S phase and the dynamics decreased significantly in S phase. Traditionally, replication sites were fixed on spatial structure of chromosomes by nuclear matrix or lamins. The Heun's results denied the traditional concepts, budding yeasts don't have lamins, and support that replication origins self-assemble and form replication foci.

By firing of replication origins, controlled spatially and temporally, the formation of replication foci is regulated. D. A. Jackson et al.(1998) revealed that neighboring origins fire simultaneously in mammalian cells. Spatial juxtaposition of replication sites brings **clustering** of replication forks. The clustering do **rescue of stalled replication forks** and favors normal progress of replication forks. Progress of replication forks is inhibited by many factors; collision with proteins or with complexes binding strongly on DNA, deficiency of dNTPs, nicks on template DNAs and so on. If replication forks stall and the remaining sequences from the stalled forks are not replicated, the daughter strands have nick obtained un-replicated sites. The un-replicated sites on one parent's strand hold the other strand together but not daughter strands. Therefore, the resulting sister chromatids cannot separate from each other and cannot divide into 2 daughter cells. When neighboring origins fire and a fork from one origin is stalled, fork from other origin access on an opposite direction of the stalled fork and duplicate the un-replicated sites. As other mechanism of the rescue there is application of **dormant replication origins** that excess origins don't fire in normal DNA replication.

Bacteria

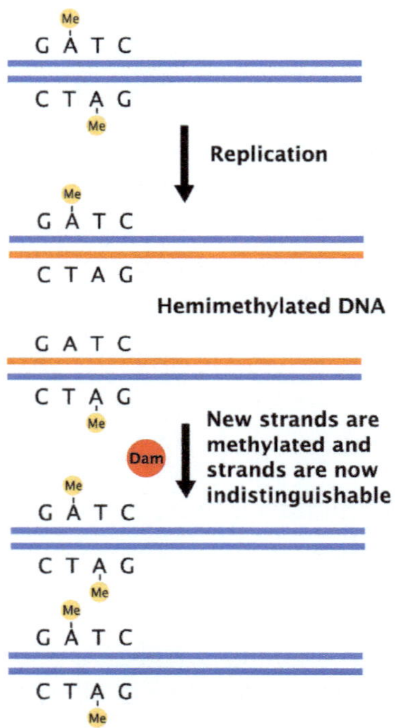

Dam methylates adenine of GATC sites after replication.

Most bacteria do not go through a well-defined cell cycle but instead continuously copy their DNA; during rapid growth, this can result in the concurrent occurrence of multiple rounds of replication. In *E. coli*, the best-characterized bacteria, DNA replication is regulated through several mechanisms, including: the hemimethylation and sequestering of the origin sequence, the ratio of adenosine triphosphate (ATP) to adenosine diphosphate (ADP), and the levels of protein DnaA. All these control the binding of initiator proteins to the origin sequences.

Because *E. coli* methylates GATC DNA sequences, DNA synthesis results in hemimethylated sequences. This hemimethylated DNA is recognized by the protein SeqA, which binds and sequesters the origin sequence; in addition, DnaA (required for initiation of replication) binds less well to hemimethylated DNA. As a result, newly replicated origins are prevented from immediately initiating another round of DNA replication.

ATP builds up when the cell is in a rich medium, triggering DNA replication once the cell has reached a specific size. ATP competes with ADP to bind to DnaA, and the DnaA-ATP complex is able to initiate replication. A certain number of DnaA proteins are also required for DNA replication — each time the origin is copied, the number of binding sites for DnaA doubles, requiring the synthesis of more DnaA to enable another initiation of replication.

Polymerase Chain Reaction

Researchers commonly replicate DNA *in vitro* using the polymerase chain reaction (PCR). PCR uses a pair of primers to span a target region in template DNA, and then polymerizes partner strands in each direction from these primers using a thermostable DNA polymerase. Repeating this process through multiple cycles amplifies the targeted DNA region. At the start of each cycle, the mixture of template and primers is heated, separating the newly synthesized molecule and template. Then, as the mixture cools, both of these become templates for annealing of new primers, and the polymerase extends from these. As a result, the number of copies of the target region doubles each round, increasing exponentially.

References

- Cooper GM (2000). "Chapter 14: The Eukaryotic Cell Cycle". The cell: a molecular approach (2nd ed.). Washington, D.C: ASM Press. ISBN 0-87893-106-6.
- Robbins, Stanley L; Cotran, Ramzi S (2004). Vinay Kumar; Abul K Abbas; Nelson Fausto, eds. Pathological Basis of Disease. Elsevier. ISBN 81-8147-528-3.
- Norbury C (1995). "Cdk2 protein kinase (vertebrates)". In Hardie, D. Grahame; Hanks, Steven. Protein kinase factsBook. Boston: Academic Press. p. 184. ISBN 0-12-324719-5.
- Maton, A.; Hopkins, J. J.; LaHart, S; Quon Warner, D.; Wright, M.; Jill, D. (1997). Cells: Building Blocks of Life. New Jersey: Prentice Hall. pp. 70–4. ISBN 0-13-423476-6.
- Olson, Mark O. J. (2011). The Nucleolus. Volume 15 of Protein Reviews. Berlin: Springer Science & Business Media. p. 15. ISBN 9781461405146.
- Iourov, I.Y.; Vorsanova, S.G.; Yurov, Y.B. (2006). "Chromosomal Variations in Mammalian Neuronal Cells: Known Facts and Attractive Hypotheses". In Jeon, Kwang J. International Review Of Cytology: A Survey of Cell Biology. 249. Waltham, MA: Academic Press. p. 146. ISBN 9780080463506.
- Robert.S Hine, ed. (2008). Oxford Dictionary Biology (6th ed.). New York: Oxford University Press. p. 113. ISBN 978-0-19-920462-5. .
- Kumar, Rani (2008). Textbook of Human Embryology. I.K. International Publishing House. p. 22. ISBN 9788190675710.
- D. Binder, Marc; Hirokawa, Nobutaka; Windhorst, Uwe (2009). Encyclopedia of Neuroscience. Springer. ISBN 3540237356.
- Stocum DL (2004). "Amphibian regeneration and stem cells". Curr. Top. Microbiol. Immunol. Current Topics in Microbiology and Immunology. 280: 1–70. doi:10.1007/978-3-642-18846-6_1. ISBN 978-3-540-02238-1. PMID 14594207.
- Knisely, Karen; Gilbert, Scott F. (2009). Developmental Biology (8th ed.). Sunderland, Mass: Sinauer Associates. p. 147. ISBN 0-87893-371-9.

- Berg JM, Tymoczko JL, Stryer L, Clarke ND (2002). "Chapter 27, Section 4: DNA Replication of Both Strands Proceeds Rapidly from Specific Start Sites". Biochemistry. W.H. Freeman and Company. ISBN 0-7167-3051-0. External link in |chapter= (help)
- Alberts B, Johnson A, Lewis J, Raff M, Roberts K, Walter P (2002). Molecular Biology of the Cell. Garland Science. ISBN 0-8153-3218-1.
- Lodish H, Berk A, Zipursky LS, Matsudaira P, Baltimore D, Darnell J (2000). Molecular Cell Biology. W. H. Freeman and Company. ISBN 0-7167-3136-3. 12.1.
- Alberts B, Johnson A, Lewis J, Raff M, Roberts K, Walter P (2002). Molecular Biology of the Cell. Garland Science. ISBN 0-8153-3218-1.
- Griffiths, Anthony J.F.; Wessler, Susan R.; Carroll, Sean B.; Doebley, John (2012). Introduction to Genetic Analysis (10 ed.). New York: W.H. Freeman and Company. p. 35. ISBN 978-1-4292-2943-2.

6

Integrated Fields of Cell Biology

Cell function and cell metabolism require certain fundamental activities to survive such as the ingestion and digestion of food and the flow of blood, in the case of animals. Cells also perform complex actions that can be defined under cell adhesion.

Cell Physiology

Cell physiology (including cellular electrophysiology) is the biological study of the activities which take place in a cell to keep it alive. The term "physiology" refers to all the normal functions that take place in a living organism. Absorption of water by roots, production of food in the leaves, and growth of shoots towards light are examples of plant physiology. The heterotrophic metabolism of food derived from plants and animals and the use of movement to obtain nutrients (even if the organism itself remains in a relatively stationary position) are characteristic of animal physiology.

In the context of human physiology, the term *cell physiology* often specifically applies to the physiology of membrane transport, neuron transmission, and (less frequently) muscle contraction. In general, these cover the digestion of food, circulation of blood, and contraction of muscles and, therefore, are important aspects of human physiology. For a more complete description of the general physiological function of human cells (as well as the cells of other life forms).

General Characteristics of Cell Physiology

There are two types of cells, Prokaryotes and Eukaryotes.

Prokaryotes first came into existence and contain no self-contained nucleus, therefore making their mechanisms much simpler compared to the later-evolved Eukaryotes, which do contain a nucleus enveloping the cell's DNA and nuclear organelles. Because viruses, viroids, prions and such depend entirely on the physiology of other cells (i.e., cells containing their own physiology), the former entities are often not considered to be "living" by the biologists who study them.

All living cells, whether prokaryotes or eukaryotes, contain the following distinguishing characteristics:

Based on the properties shared by all independently living organisms on Earth,

- The genetic code is based on DNA.
    - The DNA is composed of four nucleotides (deoxyadenosine, deoxycytidine, deoxy-

thymidine and deoxyguanosine), to the exclusion of other possible deoxynucleotides.

- The genetic code is composed of three-nucleotide codons, thus producing 64 different codons. Since only 20 amino acids are used, multiple codons code for the same amino acids. This structure is arbitrary and shared by all eukaryotes and prokaryotes. Archaea and mitochondria use a similar code with minor differences.
- The DNA is kept double-stranded by a template-dependent DNA polymerase.
- The integrity of the DNA is maintained by a group of maintenance enzymes, including DNA topoisomerase, DNA ligase and other DNA repair enzymes. The DNA is also protected by DNA-binding proteins like histones.

- The genetic code is expressed via RNA intermediates, which are single-stranded.
  - RNA is produced by a DNA-dependent RNA polymerase using nucleotides similar to those of DNA with the exception of Thymidine in DNA, replaced by Uridine in RNA.
- The genetic code is expressed into proteins. All other properties of the organism (e.g. synthesis of lipids or carbohydrates) are the result of protein enzymes.
- Proteins are assembled from free amino acids by translation of an mRNA by ribosomes, tRNA and a group of related proteins.
  - Ribosomes are composed of two subunits, one big and one small.
  - Each ribosomal subunit is composed of a core of ribosomal RNA surrounded by ribosomal proteins.
  - The RNA molecules (rRNA and tRNA) play an important role in the catalytic activity of the ribosomes
- Only 20 amino acids are used, to the exclusion of countless non-standard amino acids; only the L-isomers are used.
  - Amino acids must be synthesized from glucose by a group of specialized enzymes; the synthesis pathways are arbitrary and conserved.
- Glucose can be used as a source of energy and carbon; only the D-isomer is used.
  - Glycolysis goes through an arbitrary degradation pathway.
- ATP is used as an energy intermediate.
- The cell is surrounded by a cellular membrane composed of a lipid bilayer.
- Inside the cell, the concentration of sodium is lower, and potassium is higher, than outside. This gradient is maintained by specific ion pumps.The concentration of Calcium inside of a cell is also lower than outside.
- The cell multiplies by duplicating all its contents followed by cellular division.

The earliest ancestor of all life that is hypothesized to contain these attributes is known as the last common ancestor.

Physiological Processes

- Movement of proteins — The movement of proteins about the cell for use in structure and in enzymatic processes.
- Active transport and Passive transport — The processes facilitating the movement of molecules into and out of cells.
- Autophagy — The process whereby cells "eat" their own internal components or microbial invaders.
- Adhesion — The chemical processes whereby cells and other tissues are held together.
- Cell division — a eukaryotic cell process resulting in the formation of daughter cells; there are two major types: that of mitosis (asexual reproduction) and meiosis (sexual reproduction).
- Cell movement: Chemotaxis, Contraction, Cilia and Flagella.
- Cell signaling — Regulation of cell behavior by signals from outside, such as the use of hormones or neurotransmitters.
- DNA repair and Cell death
- Metabolism: Glycolysis, Respiration, Photosynthesis — Processes whereby energy is stored and/or liberated for use by the cell.
- Transcription and mRNA splicing — Processes by which genes express themselves, primarily by way of RNA and protein transcription.

Cell Adhesion

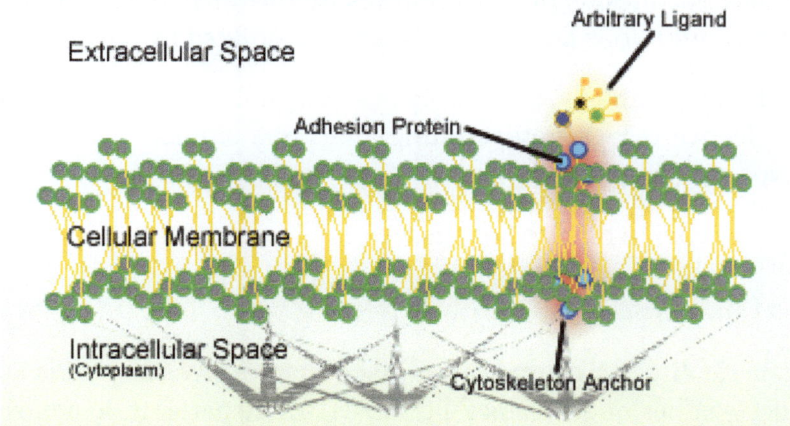

Schematic of cell adhesion

Cell adhesion is the process cells interact and attach to a surface, substrate or another cell, mediated by interactions between molecules of the cell surface. Cell adhesion occurs from the action of transmembrane glycoproteins, called cell adhesion molecules. Examples of these proteins include selectins, integrins, syndecans, and cadherins. Cellular adhesion is essential in maintaining multicellular structure. Cellular adhesion can link cells in different ways and can be involved in signal transduction. Cell adhesion is also essential for the pathogenesis of infectious organisms.

Process

Eukaryotes

Protozoans express multiple adhesion molecules. An example of a pathogenic protozoan is the malarial parasite (*Plasmodium falciparum*), which uses one adhesion molecule called the circumsporozoite protein to bind to liver cells, and another adhesion molecule called the merozoite surface protein to bind red blood cells.

Mammalian cells express many different types of adhesion molecules. The major classes are named integrins, Ig superfamily members, cadherins, and selectins. Each of these adhesion molecules has a different function and recognizes different ligands. Defects in cell adhesion are usually attributable to defects in expression of adhesion molecules.

Cell junctions allow cells to adhere to one another in multicellular organisms. There are 4 types of cell junctions:

- Anchoring junctions (Adherens junctions, Desmosomes and Hemidesmosomes), which transmit stress through tethering to cytoskeleton are cell–cell or cell–matrix
- Occluding Junctions (Tight junctions), which seal gaps between cells, making an impermeable barrier
- Channel forming junctions (Gap junctions and plasmodesmata), which links cytoplasm of adjacent cells
- Signal relaying junction, synapses in nervous systems

Cell-cell adhesions in Anchoring junctions are mediated by cadherins. The cadherin subunit, β-catenin plays a role in cell-cell adhesion by controlling cadherin-mediated cell adhesion at the plasma membrane. Cell–matrix adhesions however, are usually mediated by integrins.

Selective cell–cell adhesion enables vertebrate cells to assemble into organised tissues. homophilic attachment allows selective recognition resulting. Cells of a similar type stick together whereas cells of a different type stay segregated.

Fungal adhesion allows fungi to bind to host tissues.

Human Genetic Diseases

There are human genetic diseases caused by an inability to express a specific adhesion molecule. An example is leukocyte adhesion deficiency-I (LAD-I), where patients do not express the β2-integrin subunit precursor. This integrin is required for leukocytes to adhere to the blood vessel wall during inflammation in order to fight infection. The leukocytes from LAD-I patients fail to adhere and patients exhibit serious episodes of infection that can be life-threatening.

Tumor Adhesion

Cancer metastasis tumors that spread through the circulatory system use mechanisms of cell adhesion to establish new tumors in the body. Release of epoxyeicosatrienoic acid is believed to increase this propensity.

Prokaryotes

Prokaryotes have adhesion molecules usually termed "adhesins". Adhesins may occur on pili (fimbriae), flagellae, or the cell surface. Adhesion of bacteria is the first step in colonization and regulates tropism (tissue- or cell-specific interactions).

Viruses

Viruses also have adhesion molecules required for viral binding to host cells. For example, influenza virus has a hemagglutinin on its surface that is required for recognition of the sugar sialic acid on host cell surface molecules. HIV has an adhesion molecule termed gp120 that binds to its ligand CD4, which is expressed on lymphocytes.

Differential Adhesion Hypothesis

The differential adhesion hypothesis (sometimes called the "thermodynamic hypothesis") is a theory of cell adhesion advanced by Malcolm Steinberg in 1964 to explain the mechanism by which heterotypic cells in mixed aggregates sort out into isotypic territories. The DAH postulates that tissues are viscoelastic liquids, and as such possess measurable tissue surface tensions. These surface tensions have been determined for a variety of tissues, including embryonic tissues and cell lines. The surface tensions correspond to the mutual sorting behavior: the tissue type with the higher surface tension will occupy an internal position relative to a tissue with a lower surface tension (if these tissues can interact with each other through their adhesion machinery).

Quantitative differences in homo and heterotypic adhesion are supposed to be sufficient to account for the phenomenon without the need to postulate cell type specific adhesion systems: fairly generally accepted, although some tissue specific cell adhesion molecules are now known to exist.

Clinical Implications

Dysfunction of cell-adhesion and cell-migration occurs during cancer metastasis. Cellular adhesion and traction can allow cells to migrate. Cells can form integrin mediated attachments sites called focal adhesions. Focal adhesions at the leading edge provide the necessary traction allowing the cell to pull itself forward.

Pemphigus is the result of auto-antibodies which target desmosomal cadherins, resulting in loss of cell adhesion.

Measurement and Applications

Cell adhesion forces of mammalian cells could specifically be measured down to the single receptor-ligand level. Researchers at Japan's Nagoya University have developed a test to measure the adhesion of a single cell and potentially determine cellular viability.

Cell (Biology)

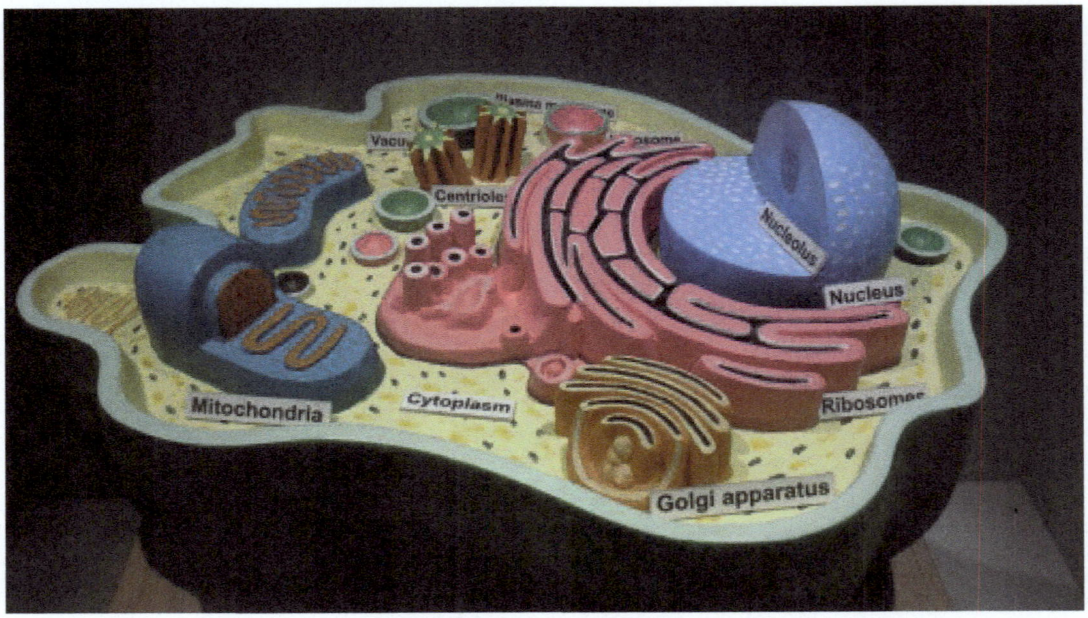

Structure of an animal cell

The cell (from Latin *cella*, meaning "small room") is the basic structural, functional, and biological unit of all known living organisms. A cell is the smallest unit of life that can replicate independently, and cells are often called the "building blocks of life". The study of cells is called cell biology.

Cells consist of cytoplasm enclosed within a membrane, which contains many biomolecules such as proteins and nucleic acids. Organisms can be classified as unicellular (consisting of a single cell; including bacteria) or multicellular (including plants and animals). While the number of cells in plants and animals varies from species to species, humans contain more than 10 trillion (10^{13}) cells. Most plant and animal cells are visible only under a microscope, with dimensions between 1 and 100 micrometres.

The cell was discovered by Robert Hooke in 1665, who named the biological unit for its resemblance to cells inhabited by Christian monks in a monastery. Cell theory, first developed in 1839 by Matthias Jakob Schleiden and Theodor Schwann, states that all organisms are composed of one or more cells, that cells are the fundamental unit of structure and function in all living organisms, that all cells come from preexisting cells, and that all cells contain the hereditary information necessary for regulating cell functions and for transmitting information to the next generation of cells. Cells emerged on Earth at least 3.5 billion years ago.

Anatomy

Comparison of Features of Prokaryotic and Eukaryotic Cells		
	Prokaryotes	**Eukaryotes**
Typical organisms	bacteria, archaea	protists, fungi, plants, animals
Typical size	~ 1–5 µm	~ 10–100 µm
Type of nucleus	nucleoid region; no true nucleus	true nucleus with double membrane
DNA	circular (usually)	linear molecules (chromosomes) with histone proteins
RNA/protein synthesis	coupled in the cytoplasm	RNA synthesis in the nucleus protein synthesis in the cytoplasm
Ribosomes	50S and 30S	60S and 40S
Cytoplasmic structure	very few structures	highly structured by endomembranes and a cytoskeleton
Cell movement	flagella made of flagellin	flagella and cilia containing microtubules; lamellipodia and filopodia containing actin
Mitochondria	none	one to several thousand
Chloroplasts	none	in algae and plants
Organization	usually single cells	single cells, colonies, higher multicellular organisms with specialized cells
Cell division	binary fission (simple division)	mitosis (fission or budding) meiosis
Chromosomes	single chromosome	more than one chromosome
Membranes	cell membrane	Cell membrane and membrane-bound organelles

Cells are of two types, eukaryotic, which contain a nucleus, and prokaryotic, which do not. Prokaryotes are single-celled organisms, while eukaryotes can be either single-celled or multicellular.

Prokaryotic Cells

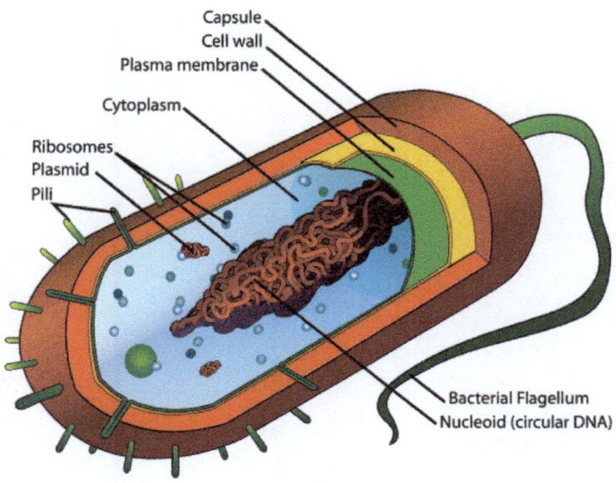

Structure of a typical prokaryotic cell

Prokaryotic cells were the first form of life on Earth, characterised by having vital biological processes including cell signaling and being self-sustaining. They are simpler and smaller than eukaryotic cells, and lack membrane-bound organelles such as the nucleus. Prokaryotes include two of the domains of life, bacteria and archaea. The DNA of a prokaryotic cell consists of a single chromosome that is in direct contact with the cytoplasm. The nuclear region in the cytoplasm is called the nucleoid. Most prokaryotes are the smallest of all organisms ranging from 0.5 to 2.0 μm in diameter.

A prokaryotic cell has three architectural regions:

- Enclosing the cell is the cell envelope – generally consisting of a plasma membrane covered by a cell wall which, for some bacteria, may be further covered by a third layer called a capsule. Though most prokaryotes have both a cell membrane and a cell wall, there are exceptions such as *Mycoplasma* (bacteria) and *Thermoplasma* (archaea) which only possess the cell membrane layer. The envelope gives rigidity to the cell and separates the interior of the cell from its environment, serving as a protective filter. The cell wall consists of peptidoglycan in bacteria, and acts as an additional barrier against exterior forces. It also prevents the cell from expanding and bursting (cytolysis) from osmotic pressure due to a hypotonic environment. Some eukaryotic cells (plant cells and fungal cells) also have a cell wall.
- Inside the cell is the cytoplasmic region that contains the genome (DNA), ribosomes and various sorts of inclusions. The genetic material is freely found in the cytoplasm. Prokaryotes can carry extrachromosomal DNA elements called plasmids, which are usually circular. Linear bacterial plasmids have been identified in several species of spirochete bacteria, including members of the genus Borrelia notably *Borrelia burgdorferi*, which causes Lyme disease. Though not forming a *nucleus*, the DNA is condensed in a *nucleoid*. Plasmids encode additional genes, such as antibiotic resistance genes.
- On the outside, flagella and pili project from the cell's surface. These are structures (not present in all prokaryotes) made of proteins that facilitate movement and communication between cells.

Eukaryotic Cells

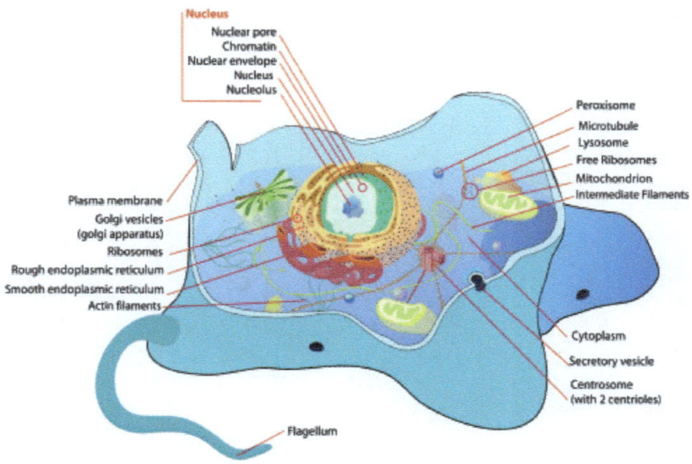

Structure of a typical animal cell

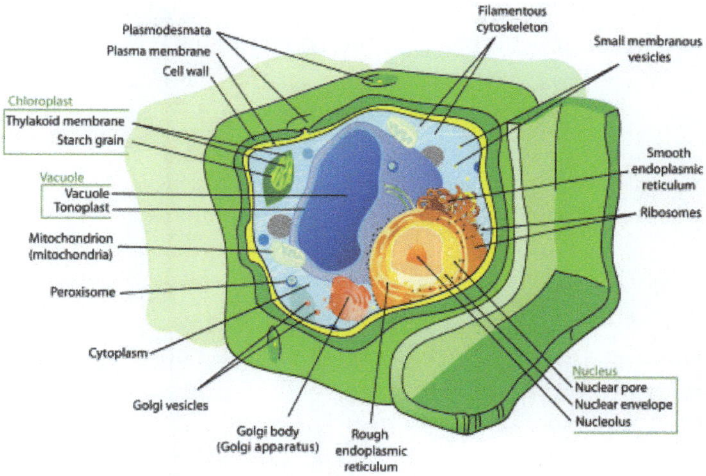

Structure of a typical plant cell

Plants, animals, fungi, slime moulds, protozoa, and algae are all eukaryotic. These cells are about fifteen times wider than a typical prokaryote and can be as much as a thousand times greater in volume. The main distinguishing feature of eukaryotes as compared to prokaryotes is compartmentalization: the presence of membrane-bound organelles (compartments) in which specific metabolic activities take place. Most important among these is a cell nucleus, an organelle that houses the cell's DNA. This nucleus gives the eukaryote its name, which means "true kernel (nucleus)". Other differences include:

- The plasma membrane resembles that of prokaryotes in function, with minor differences in the setup. Cell walls may or may not be present.
- The eukaryotic DNA is organized in one or more linear molecules, called chromosomes, which are associated with histone proteins. All chromosomal DNA is stored in the *cell nucleus*, separated from the cytoplasm by a membrane. Some eukaryotic organelles such as mitochondria also contain some DNA.
- Many eukaryotic cells are ciliated with *primary cilia*. Primary cilia play important roles in chemosensation, mechanosensation, and thermosensation. Cilia may thus be "viewed as a sensory cellular antennae that coordinates a large number of cellular signaling pathways, sometimes coupling the signaling to ciliary motility or alternatively to cell division and differentiation."
- Motile cells of eukaryotes can move using *motile cilia* or *flagella*. Motile cells are absent in conifers and flowering plants. Eukaryotic flagella are less complex than those of prokaryotes.

Subcellular Components

All cells, whether prokaryotic or eukaryotic, have a membrane that envelops the cell, regulates what moves in and out (selectively permeable), and maintains the electric potential of the cell. Inside the membrane, the cytoplasm takes up most of the cell's volume. All cells (except red blood cells which lack a cell nucleus and most organelles to accommodate maximum space for hemoglobin) possess DNA, the hereditary material of genes, and RNA, containing the information necessary to build various proteins such as enzymes, the cell's primary machinery. There are also other kinds of biomolecules in cells. This article lists these primary components of the cell, then briefly describes their function.

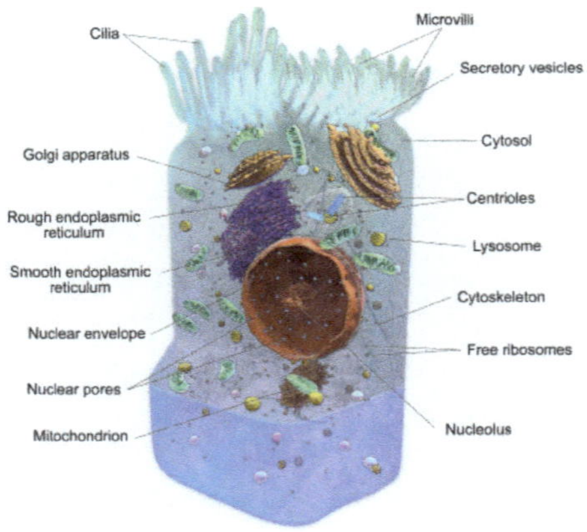

Anatomy of a Cell

Illustration depicting major structures inside a eukaryotic animal cell

Membrane

The cell membrane, or plasma membrane, is a biological membrane that surrounds the cytoplasm of a cell. In animals, the plasma membrane is the outer boundary of the cell, while in plants and prokaryotes it is usually covered by a cell wall. This membrane serves to separate and protect a cell from its surrounding environment and is made mostly from a double layer of phospholipids, which are amphiphilic (partly hydrophobic and partly hydrophilic). Hence, the layer is called a phospholipid bilayer, or sometimes a fluid mosaic membrane. Embedded within this membrane is a variety of protein molecules that act as channels and pumps that move different molecules into and out of the cell. The membrane is said to be 'semi-permeable', in that it can either let a substance (molecule or ion) pass through freely, pass through to a limited extent or not pass through at all. Cell surface membranes also contain receptor proteins that allow cells to detect external signaling molecules such as hormones.

Cytoskeleton

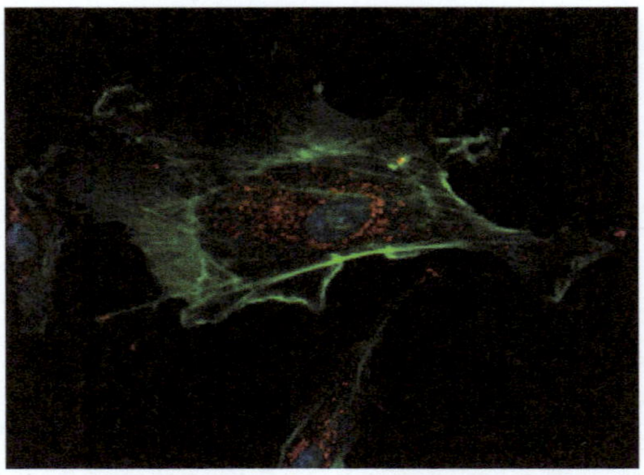

A fluorescent image of an endothelial cell. Nuclei are stained blue, mitochondria are stained red, and microfilaments are stained green.

The cytoskeleton acts to organize and maintain the cell's shape; anchors organelles in place; helps during endocytosis, the uptake of external materials by a cell, and cytokinesis, the separation of daughter cells after cell division; and moves parts of the cell in processes of growth and mobility. The eukaryotic cytoskeleton is composed of microfilaments, intermediate filaments and microtubules. There are a great number of proteins associated with them, each controlling a cell's structure by directing, bundling, and aligning filaments. The prokaryotic cytoskeleton is less well-studied but is involved in the maintenance of cell shape, polarity and cytokinesis. The subunit protein of microfilaments is a small, monomeric protein called actin. The subunit of microtubules is a dimeric molecule called tubulin. Intermediate filaments are heteropolymers whose subunits vary among the cell types in different tissues. But some of the subunit protein of intermediate filaments include vimentin, desmin, lamin (lamins A, B and C), keratin (multiple acidic and basic keratins), neurofilament proteins (NF - L, NF - M).

Genetic Material

Two different kinds of genetic material exist: deoxyribonucleic acid (DNA) and ribonucleic acid (RNA). Cells use DNA for their long-term information storage. The biological information contained in an organism is encoded in its DNA sequence. RNA is used for information transport (e.g., mRNA) and enzymatic functions (e.g., ribosomal RNA). Transfer RNA (tRNA) molecules are used to add amino acids during protein translation.

Prokaryotic genetic material is organized in a simple circular DNA molecule (the bacterial chromosome) in the nucleoid region of the cytoplasm. Eukaryotic genetic material is divided into different, linear molecules called chromosomes inside a discrete nucleus, usually with additional ge-netic material in some organelles like mitochondria and chloroplasts.

A human cell has genetic material contained in the cell nucleus (the nuclear genome) and in the mitochondria (the mitochondrial genome). In humans the nuclear genome is divided into 46 linear DNA molecules called chromosomes, including 22 homologous chromosome pairs and a pair of sex chromosomes. The mitochondrial genome is a circular DNA molecule distinct from the nuclear DNA. Although the mitochondrial DNA is very small compared to nuclear chromosomes, it codes for 13 proteins involved in mitochondrial energy production and specific tRNAs.

Foreign genetic material (most commonly DNA) can also be artificially introduced into the cell by a process called transfection. This can be transient, if the DNA is not inserted into the cell's genome, or stable, if it is. Certain viruses also insert their genetic material into the genome.

Organelles

Organelles are parts of the cell which are adapted and/or specialized for carrying out one or more vital functions, analogous to the organs of the human body (such as the heart, lung, and kidney, with each organ performing a different function). Both eukaryotic and prokaryotic cells have organelles, but prokaryotic organelles are generally simpler and are not membrane-bound.

There are several types of organelles in a cell. Some (such as the nucleus and golgi apparatus) are typically solitary, while others (such as mitochondria, chloroplasts, peroxisomes and lysosomes) can be numerous (hundreds to thousands). The cytosol is the gelatinous fluid that fills the cell and surrounds the organelles.

Eukaryotic

- Cell nucleus: A cell's information center, the cell nucleus is the most conspicuous organelle found in a eukaryotic cell. It houses the cell's chromosomes, and is the place where almost all DNA replication and RNA synthesis (transcription) occur. The nucleus is spherical and separated from the cytoplasm by a double membrane called the nuclear envelope. The nuclear envelope isolates and protects a cell's DNA from various molecules that could accidentally damage its structure or interfere with its processing. During processing, DNA is transcribed, or copied into a special RNA, called messenger RNA (mRNA). This mRNA is then transported out of the nucleus, where it is translated into a specific protein molecule. The nucleolus is a specialized region within the nucleus where ribosome subunits are assembled. In prokaryotes, DNA processing takes place in the cytoplasm.
- Mitochondria and Chloroplasts: generate energy for the cell. Mitochondria are self-replicating organelles that occur in various numbers, shapes, and sizes in the cytoplasm of all eukaryotic cells. Respiration occurs in the cell mitochondria, which generate the cell's energy by oxidative phosphorylation, using oxygen to release energy stored in cellular nutrients (typically pertaining to glucose) to generate ATP. Mitochondria multiply by binary fission, like prokaryotes. Chloroplasts can only be found in plants and algae, and they capture the sun's energy to make carbohydrates through photosynthesis.

Diagram of an endomembrane system

- **Endoplasmic reticulum**: The endoplasmic reticulum (ER) is a transport network for molecules targeted for certain modifications and specific destinations, as compared to molecules that float freely in the cytoplasm. The ER has two forms: the rough ER, which has ribosomes on its surface that secrete proteins into the ER, and the smooth ER, which lacks ribosomes. The smooth ER plays a role in calcium sequestration and release.
- **Golgi apparatus**: The primary function of the Golgi apparatus is to process and package the macromolecules such as proteins and lipids that are synthesized by the cell.

- Lysosomes and Peroxisomes: Lysosomes contain digestive enzymes (acid hydrolases). They digest excess or worn-out organelles, food particles, and engulfed viruses or bacteria. Peroxisomes have enzymes that rid the cell of toxic peroxides. The cell could not house these destructive enzymes if they were not contained in a membrane-bound system.

- Centrosome: the cytoskeleton organiser: The centrosome produces the microtubules of a cell – a key component of the cytoskeleton. It directs the transport through the ER and the Golgi apparatus. Centrosomes are composed of two centrioles, which separate during cell division and help in the formation of the mitotic spindle. A single centrosome is present in the animal cells. They are also found in some fungi and algae cells.

- Vacuoles: Vacuoles sequester waste products and in plant cells store water. They are often described as liquid filled space and are surrounded by a membrane. Some cells, most notably *Amoeba*, have contractile vacuoles, which can pump water out of the cell if there is too much water. The vacuoles of plant cells and fungal cells are usually larger than those of animal cells.

Eukaryotic and Prokaryotic

- Ribosomes: The ribosome is a large complex of RNA and protein molecules. They each consist of two subunits, and act as an assembly line where RNA from the nucleus is used to synthesise proteins from amino acids. Ribosomes can be found either floating freely or bound to a membrane (the rough endoplasmatic reticulum in eukaryotes, or the cell membrane in prokaryotes).

Structures Outside the Cell Membrane

Many cells also have structures which exist wholly or partially outside the cell membrane. These structures are notable because they are not protected from the external environment by the semipermeable cell membrane. In order to assemble these structures, their components must be carried across the cell membrane by export processes.

Cell Wall

Many types of prokaryotic and eukaryotic cells have a cell wall. The cell wall acts to protect the cell mechanically and chemically from its environment, and is an additional layer of protection to the cell membrane. Different types of cell have cell walls made up of different materials; plant cell walls are primarily made up of cellulose, fungi cell walls are made up of chitin and bacteria cell walls are made up of peptidoglycan.

Prokaryotic

Capsule

A gelatinous capsule is present in some bacteria outside the cell membrane and cell wall. The capsule may be polysaccharide as in pneumococci, meningococci or polypeptide as *Bacillus anthracis* or hyaluronic acid as in streptococci. Capsules are not marked by normal staining protocols and can be detected by India ink or methyl blue; which allows for higher contrast between the cells for observation.

Flagella

Flagella are organelles for cellular mobility. The bacterial flagellum stretches from cytoplasm through the cell membrane(s) and extrudes through the cell wall. They are long and thick thread-like appendages, protein in nature. A different type of flagellum is found in archaea and a different type is found in eukaryotes.

Fimbria

A fimbria also known as a pilus is a short, thin, hair-like filament found on the surface of bacteria. Fimbriae, or pili are formed of a protein called pilin (antigenic) and are responsible for attachment of bacteria to specific receptors of human cell (cell adhesion). There are special types of specific pili involved in bacterial conjugation.

Cellular Processes

Growth and Metabolism

Between successive cell divisions, cells grow through the functioning of cellular metabolism. Cell metabolism is the process by which individual cells process nutrient molecules. Metabolism has two distinct divisions: catabolism, in which the cell breaks down complex molecules to produce energy and reducing power, and anabolism, in which the cell uses energy and reducing power to construct complex molecules and perform other biological functions. Complex sugars consumed by the organism can be broken down into simpler sugar molecules called monosaccharides such as glucose. Once inside the cell, glucose is broken down to make adenosine triphosphate (ATP), a molecule that possesses readily available energy, through two different pathways.

Replication

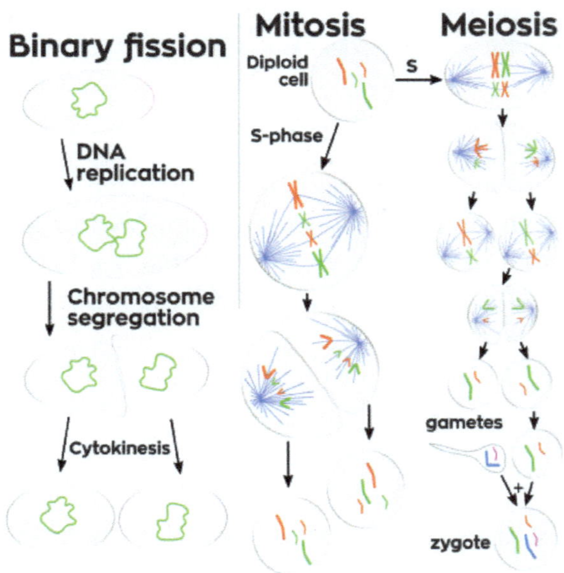

Bacteria divide by binary fission, while eukaryotes divide by mitosis or meiosis.

Cell division involves a single cell (called a *mother cell*) dividing into two daughter cells. This leads to growth in multicellular organisms (the growth of tissue) and to procreation (vegetative reproduction) in unicellular organisms. Prokaryotic cells divide by binary fission, while eukaryotic cells usually undergo a process of nuclear division, called mitosis, followed by division of the cell, called cytokinesis. A diploid cell may also undergo meiosis to produce haploid cells, usually four. Haploid cells serve as gametes in multicellular organisms, fusing to form new diploid cells.

DNA replication, or the process of duplicating a cell's genome, always happens when a cell divides through mitosis or binary fission. This occurs during the S phase of the cell cycle.

In meiosis, the DNA is replicated only once, while the cell divides twice. DNA replication only occurs before meiosis I. DNA replication does not occur when the cells divide the second time, in meiosis II. Replication, like all cellular activities, requires specialized proteins for carrying out the job.

Protein Synthesis

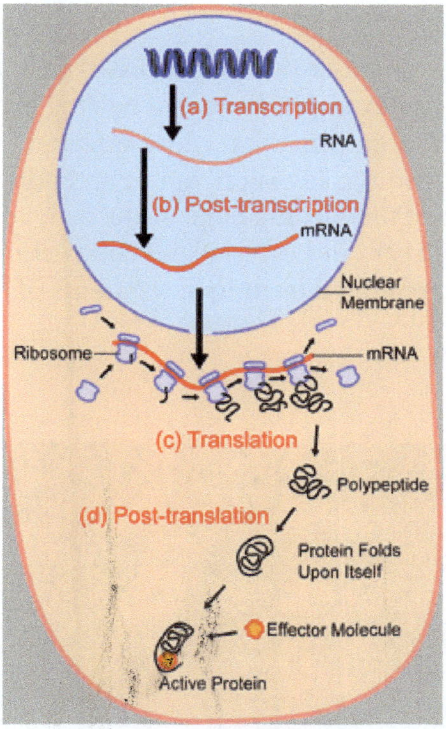

An overview of protein synthesis.

Within the nucleus of the cell (*light blue*), genes (DNA, *dark blue*) are transcribed into RNA. This RNA is then subject to post-transcriptional modification and control, resulting in a mature mRNA (*red*) that is then transported out of the nucleus and into the cytoplasm (*peach*), where it undergoes translation into a protein. mRNA is translated by ribosomes (*purple*) that match the three-base codons of the mRNA to the three-base anti-codons of the appropriate tRNA. Newly synthesized proteins (*black*) are often further modified, such as by binding to an effector molecule (*orange*), to become fully active.

Cells are capable of synthesizing new proteins, which are essential for the modulation and maintenance of cellular activities. This process involves the formation of new protein molecules from amino acid building blocks based on information encoded in DNA/RNA. Protein synthesis generally consists of two major steps: transcription and translation.

Transcription is the process where genetic information in DNA is used to produce a complementary RNA strand. This RNA strand is then processed to give messenger RNA (mRNA), which is free to migrate through the cell. mRNA molecules bind to protein-RNA complexes called ribosomes located in the cytosol, where they are translated into polypeptide sequences. The ribosome mediates the formation of a polypeptide sequence based on the mRNA sequence. The mRNA sequence directly relates to the polypeptide sequence by binding to transfer RNA (tRNA) adapter molecules in binding pockets within the ribosome. The new polypeptide then folds into a functional three-dimensional protein molecule.

Movement or Motility

Unicellular organisms can move in order to find food or escape predators. Common mechanisms of motion include flagella and cilia.

In multicellular organisms, cells can move during processes such as wound healing, the immune response and cancer metastasis. For example, in wound healing in animals, white blood cells move to the wound site to kill the microorganisms that cause infection. Cell motility involves many receptors, crosslinking, bundling, binding, adhesion, motor and other proteins. The process is divided into three steps – protrusion of the leading edge of the cell, adhesion of the leading edge and de-adhesion at the cell body and rear, and cytoskeletal contraction to pull the cell forward. Each step is driven by physical forces generated by unique segments of the cytoskeleton.

Multicellularity

Cell Specialization

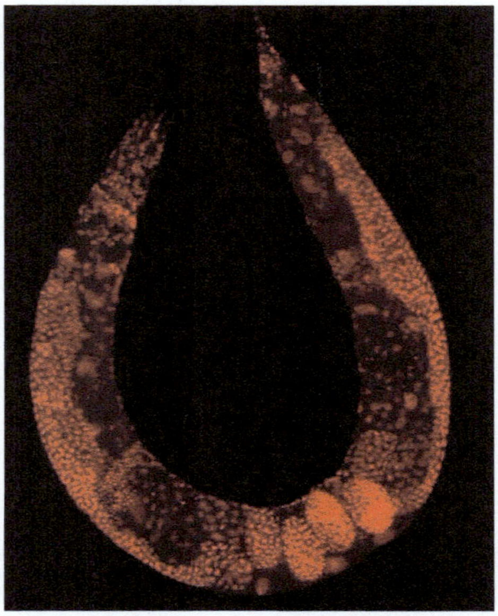

Staining of a *Caenorhabditis elegans* which highlights the nuclei of its cells.

Multicellular organisms are organisms that consist of more than one cell, in contrast to single-celled organisms.

In complex multicellular organisms, cells specialize into different cell types that are adapted to particular functions. In mammals, major cell types include skin cells, muscle cells, neurons, blood cells, fibroblasts, stem cells, and others. Cell types differ both in appearance and function, yet are genetically identical. Cells are able to be of the same genotype but of different cell type due to the differential expression of the genes they contain.

Most distinct cell types arise from a single totipotent cell, called a zygote, that differentiates into hundreds of different cell types during the course of development. Differentiation of cells is driven by different environmental cues (such as cell–cell interaction) and intrinsic differences (such as those caused by the uneven distribution of molecules during division).

Origin of Multicellularity

Multicellularity has evolved independently at least 25 times, including in some prokaryotes, like cyanobacteria, myxobacteria, actinomycetes, *Magnetoglobus multicellularis* or *Methanosarcina*. However, complex multicellular organisms evolved only in six eukaryotic groups: animals, fungi, brown algae, red algae, green algae, and plants. It evolved repeatedly for plants (Chloroplastida), once or twice for animals, once for brown algae, and perhaps several times for fungi, slime molds, and red algae. Multicellularity may have evolved from colonies of interdependent organisms, from cellularization, or from organisms in symbiotic relationships.

The first evidence of multicellularity is from cyanobacteria-like organisms that lived between 3 and 3.5 billion years ago. Other early fossils of multicellular organisms include the contested Grypania spiralis and the fossils of the black shales of the Palaeoproterozoic Francevillian Group Fossil B Formation in Gabon.

The evolution of multicellularity from unicellular ancestors has been replicated in the laboratory, in evolution experiments using predation as the selective pressure.

Origins

The origin of cells has to do with the origin of life, which began the history of life on Earth.

Origin of the First Cell

Stromatolites are left behind by cyanobacteria, also called blue-green algae. They are the oldest known fossils of life on Earth. This one-billion-year-old fossil is from Glacier National Park in the United States.

There are several theories about the origin of small molecules that led to life on the early Earth. They may have been carried to Earth on meteorites, created at deep-sea vents, or synthesized by lightning in a reducing atmosphere. There is little experimental data defining what the first self-replicating forms were. RNA is thought to be the earliest self-replicating molecule, as it is capable of both storing genetic information and catalyzing chemical reactions, but some other entity with the potential to self-replicate could have preceded RNA, such as clay or peptide nucleic acid.

Cells emerged at least 3.5 billion years ago. The current belief is that these cells were heterotrophs. The early cell membranes were probably more simple and permeable than modern ones, with only a single fatty acid chain per lipid. Lipids are known to spontaneously form bilayered vesicles in water, and could have preceded RNA, but the first cell membranes could also have been produced by catalytic RNA, or even have required structural proteins before they could form.

Origin of Eukaryotic Cells

The eukaryotic cell seems to have evolved from a symbiotic community of prokaryotic cells. DNA-bearing organelles like the mitochondria and the chloroplasts are descended from ancient symbiotic oxygen-breathing proteobacteria and cyanobacteria, respectively, which were endosymbiosed by an ancestral archaean prokaryote.

There is still considerable debate about whether organelles like the hydrogenosome predated the origin of mitochondria, or vice versa.

History of Research

- 1632–1723: Antonie van Leeuwenhoek teaches himself to make lenses, constructs basic optical microscopes and draws protozoa, such as *Vorticella* from rain water, and bacteria from his own mouth.
- 1665: Robert Hooke discovers cells in cork, then in living plant tissue using an early compound microscope. He coins the term *cell* (from Latin *cella*, meaning "small room") in his book *Micrographia* (1665).
- 1839: Theodor Schwann and Matthias Jakob Schleiden elucidate the principle that plants and animals are made of cells, concluding that cells are a common unit of structure and development, and thus founding the cell theory.
- 1855: Rudolf Virchow states that new cells come from pre-existing cells by cell division (*omnis cellula ex cellula*).
- 1859: The belief that life forms can occur spontaneously (*generatio spontanea*) is contradicted by Louis Pasteur (1822–1895) (although Francesco Redi had performed an experiment in 1668 that suggested the same conclusion).
- 1931: Ernst Ruska builds the first transmission electron microscope (TEM) at the University of Berlin. By 1935, he has built an EM with twice the resolution of a light microscope, revealing previously unresolvable organelles.

- 1953: Watson and Crick made their first announcement on the double helix structure of DNA on February 28.
- 1981: Lynn Margulis published *Symbiosis in Cell Evolution* detailing the endosymbiotic theory.

References

- Page 108 in: Walter F., PhD. Boron (2003). Medical Physiology: A Cellular And Molecular Approaoch. Elsevier/Saunders. p. 1300. ISBN 1-4160-2328-3.
- Campbell, Neil A.; Brad Williamson; Robin J. Heyden (2006). Biology: Exploring Life. Boston, Massachusetts: Pearson Prentice Hall. ISBN 0-13-250882-6.
- Karp, Gerald (19 October 2009). Cell and Molecular Biology: Concepts and Experiments. John Wiley & Sons. p. 2. ISBN 9780470483374.
- Peter Hamilton Raven; George Brooks Johnson (2002). Biology. McGraw-Hill Education. p. 68. ISBN 978-0-07-112261-0. Retrieved 7 July 2013.
- Becker, Wayne M.; et al. (2009). The world of the cell. Pearson Benjamin Cummings. p. 480. ISBN 978-0-321-55418-5.
- Bonner, John Tyler (1998). "The Origins of Multicellularity" (PDF). Integrative Biology: Issues, News, and Reviews. 1 (1): 27–36. doi:10.1002/(SICI)1520-6602(1998)1:1<27::AID-INBI4>3.0.CO;2-6. ISSN 1093-4391. Archived from the original (PDF, 0.2 MB) on March 8, 2012.
- European Bioinformatics Institute, Karyn's Genomes: Borrelia burgdorferi, part of 2can on the EBI-EMBL database. Retrieved 5 August 2012

7

Symbiogenesis: A theory Regarding Cell Evolution

There are many theories regarding the origin and evolution of cells. One of the popular ones is symbiogenesis, which regards the two main cell types, prokaryotes and eukaryotes as interacting with each other and leading to the differentiation and classification of cells into the two groups. This chapter elucidates the crucial theories and principles of symbiogenesis.

Symbiogenesis, or endosymbiotic theory, first articulated in 1910 by the Russian botanist Konstantin Mereschowsky, is one of a number of theories regard-ing the evolution of eukaryote and prokaryote organisms. It holds that the organelles distinguish-ing eukaryote cells evolved from symbiosis of individual single-celled prokaryotes, (Bacteria and Archaea). The mitochondria, plastids (for example chloroplasts), and possibly other organelles of eukaryote cells are thought to represent formerly free-living prokaryotes taken one inside the other in endosymbiosis around 1.5 billion years ago. Molecular and biochemical evidence suggest that mitochondria developed from proteobacteria (in particular, Rickettsiales, the SAR11 clade, or close relatives) and chloroplasts from cyanobacteria (in particular, nitrogen-fixing filamentous cyanobacteria).

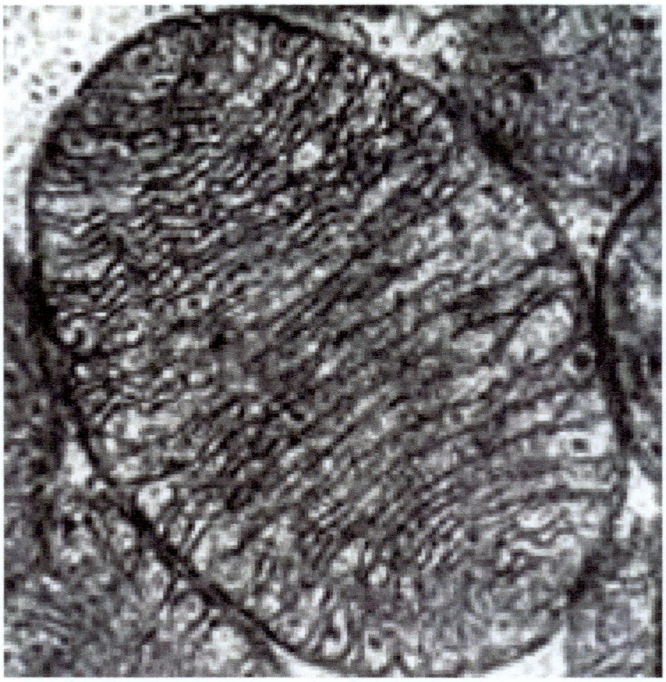

Electron micrograph of a mitochondrion showing its mitochondrial matrix and membranes

History

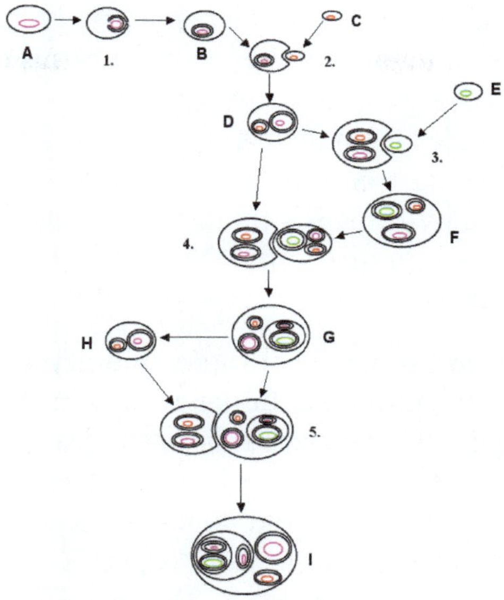

Endosymbiotic theory

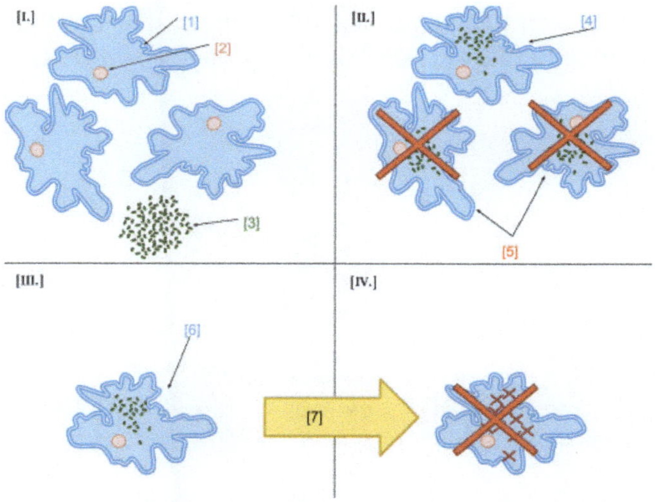

In 1966 Microbiologist Kwang Jeon conducted an experiment with amoebae communities providing real-life evidence for the endosymbiotic theory. [I.] The single-celled amoebae community was infected by a bacterial infection of x-bacteria. [II.] The x-bacteria caused many of the amoebae to become sick and die. Very few of the amoebae survived the epidemic. [III.] Amoebae surviving the epidemic have x-bacteria living inside their cytoplasm. [IV.] Antibiotics were used to kill the x-bacteria inside the remaining amoebae causing the host amoeba to die as well because the amoebae had become dependent on the x-bacteria to survive. Labels: Amoebae, Nucleos, X-Bacteria, Infected Amoebae, Sick Amoebae, Surviving Amoebae, Add Antibiotic

The theory of symbiogenesis *was first articulated by the Russian botanist Konstantin Mereschowsky in 1910, although he described the fundamental elements of the theory in a paper five years earlier. Mereschkowski was familiar with work by botanist Andreas Schimper, who had observed in 1883 that the division of chloroplasts in green plants closely resembled that of free-living cyanobacteria, and who had himself tentatively proposed (in a footnote) that green plants had arisen from a symbiotic union of two organisms. In 1918 the French scientist Paul Portier* published *Les Symbi-*

otes in which he claimed that the mitochondria originated from a symbiosis process. Ivan Wallin extended the idea of an endosymbiotic origin to mitochondria in the 1920s. The Russian botanist Boris Kozo-Polyansky was the first to explain the theory in terms of Darwinian evolution. In his 1924 book *Symbiogenesis: A New Principle of Evolution* he wrote, "The theory of symbiogenesis is a theory of selection relying on the phenomenon of symbiosis." These theories were initially dismissed or ignored. More detailed electron microscopic comparisons between cyanobacteria and chloroplasts (for example studies by Hans Ris published in 1961), combined with the discovery that plastids and mitochondria contain their own DNA (which by that stage was recognized to be the hereditary material of organisms) led to a resurrection of the idea in the 1960s.

The theory was advanced and substantiated with microbiological evidence by Lynn Margulis in a 1967 paper, *On the origin of mitosing cells*. In her 1981 work *Symbiosis in Cell Evolution* she argued that eukaryotic cells originated as communities of interacting entities, including endosymbiotic spirochaetes that developed into eukaryotic flagella and cilia. This last idea has not received much acceptance, because flagella lack DNA and do not show ultrastructural similarities to bacteria or archaea. According to Mar-gulis and Dorion Sagan, "Life did not take over the globe by combat, but by networking" (i.e., by cooperation). The possibility that peroxisomes may have an endosymbiotic origin has also been considered, although they lack DNA. Christian de Duve proposed that they may have been the first endosymbionts, allowing cells to withstand growing amounts of free molecular oxygen in the Earth's atmosphere. However, it now appears that they may be formed *de novo*, contradicting the idea that they have a symbiotic origin.

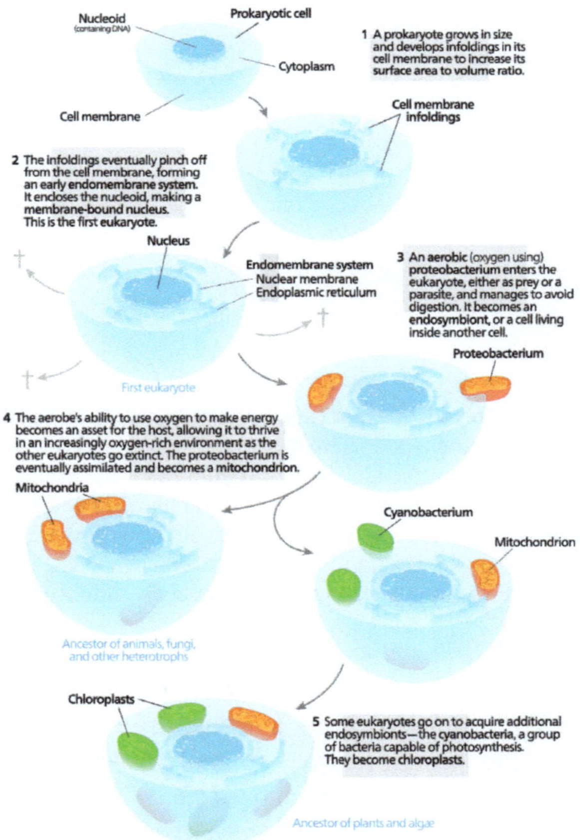

One model for the origin of mitochondria and plastids.

It is thought that over millennia these endosymbionts transferred some of their own DNA to the host cell's nucleus (called "endosymbiotic gene transfer") during the evolutionary transition from a symbiotic community to an instituted eukaryotic cell. The endosymbiotic theory is considered to be a type of saltational evolution.

From Endosymbionts to Organelles

According to Keeling and Archibald, the usual way to distinguish organelles from endosymbionts is by their reduced genome sizes. As an endosymbiont evolves into an organelle, most of their genes are transferred to the host cell genome. The host cell and organelle need to develop a transport mechanism that enables transfer back of the protein products needed by the organelle but now manufactured by the cell. Cyanobacteria and α-proteobacteria are the most closely related free-living organisms to plastids and mitochondria respectively. Both cyanobacteria and α-proteobacteria maintain a large (>6Mb) genome encoding thousands of proteins. Plastids and mitochondria exhibit a dramatic reduction in genome size when compared to their bacterial relatives. Chloroplast genomes in photosynthetic organisms are normally 120-200kb encoding 20-200 proteins and mitochondrial genomes in humans are approximately 16kb and encode 37 genes, 13 of which are proteins. Using the example of the freshwater amoeboid, however, *Paulinella chromatophora*, which contains chromatophores found to be evolved from cyanobacteria, Keeling and Archibald argue that this is not the only possible criterion; another is that the host cell has assumed control of the regulation of the former endosymbiont's division, thereby synchronizing it with the cell's own division. Nowack and her colleagues performed gene sequencing on the chromatophore (1.02 Mb) and found that only 867 proteins were encoded by these photosynthetic cells. Comparisons with their closest free living cyanobacteria of the genus *Synechococcus* (having a genome size 3 Mb, with 3300 genes) revealed that chromatophores underwent a drastic genome shrinkage. Chromatophores contained genes that were accountable for photosynthesis but were deficient in genes that could carry out other biosynthetic functions; this observation suggests that these endosymbiotic cells are highly dependent on their hosts for their survival and growth mechanisms. Thus, these chromatophores were found to be non-functional for organelle-specific purposes when compared to mitochondria and plastids. This distinction could have promoted the early evolution of photosynthetic organelles.

The loss of genetic autonomy, that is, the loss of many genes from endosymbionts, occurred very early in evolutionary time. Taking into account the entire original endosymbiont genome, there are three main possible fates for genes over evolutionary time. The first fate involves the loss of functionally redundant genes, in which genes that are already represented in the nucleus are eventually lost.The second fate involves the transfer of genes to the nucleus. The loss of autonomy and integration of the endosymbiont with its host can be primarily attributed to nuclear gene transfer. As organelle genomes have been greatly reduced over evolutionary time, nuclear genomes have expanded and become more complex. As a result, many plastid and mitochondrial processes are driven by nuclear encoded gene products. In addition, many nuclear genes originating from endosymbionts have acquired novel functions unrelated to their organelles.

The mechanisms of gene transfer are not fully known; however, multiple hypotheses exist to explain this phenomenon. The cDNA hypothesis involves the use of messenger RNA (mRNAs) to transport genes from organelles to the nucleus where they are converted to cDNA and incorporated into the genome. The cDNA hypothesis is based on studies of the genomes of flowering plants. Protein coding RNAs in mitochondria are spliced and edited using organelle-specific splice and editing sites. Nuclear copies of some mitochondrial genes, however, do not contain organelle-specific splice sites, suggesting a processed mRNA intermediate. The cDNA hypothesis has since been revised as edited mitochondrial cDNAs are unlikely to recombine with the nuclear genome and are more likely to recombine with their native mitochondrial genome. If the edited mitochondrial sequence recombines with the mitochondrial genome, mitochondrial splice sites would no longer exist in the mitochondrial genome. Any subsequent nuclear gene transfer would therefore also lack mitochondrial splice sites.

The bulk flow hypothesis is the alternative to the cDNA hypothesis, stating that escaped DNA, rather than mRNA, is the mechanism of gene transfer. According to this hypothesis, disturbances to organelles, including autophagy (normal cell destruction), gametogenesis (the formation of gametes), and cell stress, release DNA which is imported into the nucleus and incorporated into the nuclear DNA using non-homologous end joining (repair of double stranded breaks). For example, in the initial stages of endosymbiosis, due to a lack of major gene transfer, the host cell had little to no control over the endosymbiont. The endosymbiont underwent cell division independently of the host cell, resulting in many "copies" of the endosymbiont within the host cell. Some of the endosymbionts lysed (burst), and high levels of DNA were incorporated into the nucleus. A similar mechanism is thought to occur in tobacco plants, who show a high rate of gene transfer and whose cells contain multiple chloroplasts. In addition, the bulk flow hypothesis is also supported by the presence of non-random clusters of organelle genes, suggesting the simultaneous movement of multiple genes.

Organellar Genomes

Plastomes and Mitogenomes

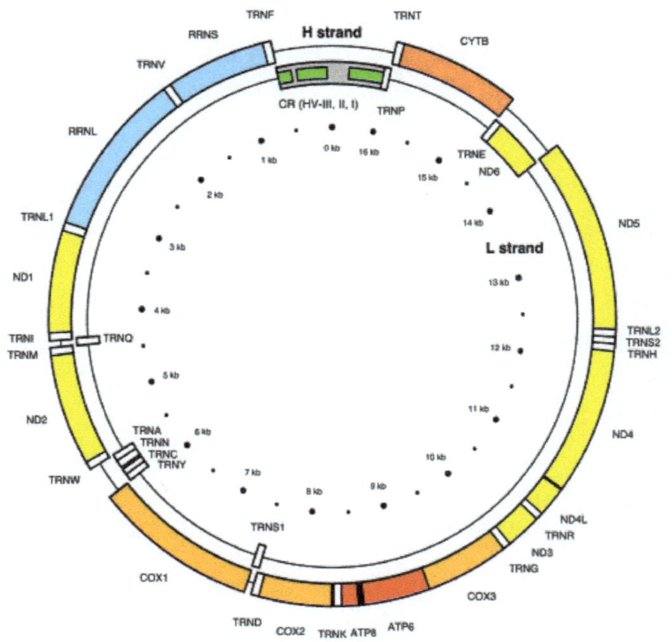

The human mitochondrial genome retained genes encoding 2 rRNAs, 22 tRNAs, and 13 redox proteins

The third and final possible fate of endosymbiont genes is that they remain in the organelles. Plastids and mitochondria, although they have lost much of their genomes, retain genes encoding rRNAs, tRNAs, proteins involved in redox reactions and proteins required for transcription, translation and replication. There are many hypotheses to explain why the organelles retain a small portion of their genome, however no one hypothesis will apply to all organisms and the topic is still quite controversial. The hydrophobicity hypothesis states that highly hydrophobic (water hating) proteins (such as the membrane bound proteins involved in redox reactions) are not easily transported through the cytosol and therefore these proteins must be encoded in their respective organelles. The code disparity hypothesis states that the limit on transfer is due to differing genetic codes and RNA editing between the organelle and the nucleus. The redox control hypothesis states that genes encoding redox reaction proteins are retained in order to effectively couple the need for repair and the synthesis of these proteins. For example, if one of the photosystems is lost from the plastid, the intermediate electron carriers may lose or gain too many electrons, signalling the need for repair of a photosystem. The time delay involved in signalling the nucleus and transporting a cytosolic protein to the organelle results in the production of damaging reactive oxygen species. The final hypothesis states that the assembly of membrane proteins, particularly those involved in redox reactions, require coordinated synthesis and assembly of subunits, however translation and protein transport coordination is more difficult to control in the cytoplasm.

Non-photosynthetic Plastid Genomes

The majority of the genes in the mitochondria and plastid are related to the expression (transcription, translation and replication) of genes encoding proteins involved in either photosynthesis (in plastids) or cellular respiration (in mitochondria). One might predict, that the loss of photosynthesis or cellular respiration would allow for the complete loss of the plastid genome or the mitochondrial genome respectively. While there are numerous examples of mitochondrial descendants (mitosomes and hydrogenosomes) that have lost their entire organellar genome, non-photosynthetic plastids tend to retain a small genome. There are two main hypotheses to explain this occurrence:

The essential tRNA hypothesis: There have been no documented functional plastid to nucleus gene transfers of genes encoding RNA products (tRNAs and rRNAs). As a result, plastids must make their own functional RNAs or import nuclear counterparts. The genes encoding tRNA-Glu and tRNA-fmet, however, appear to be indispensable. The plastid is responsible for haem biosynthesis, which requires plastid encoded tRNA-Glu (from the gene trnE) as a precursor molecule. Like other genes encoding RNAs, trnE cannot be transferred to the nucleus. In addition, it is unlikely trnE could be replaced by a cytosolic tRNA-Glu as trnE is highly conserved; single base changes in trnE have resulted in the loss of haem synthesis. The gene for tRNA-formylmethionine (tRNA-fmet) is also encoded in the plastid genome and is required for translation initiation in both plastids and mitochondria. A plastid is required to continue expressing the gene for tRNA-fmet so long as the mitochondrion is translating proteins.

The limited window hypothesis: This hypothesis offers a more general explanation for the retention of genes in non-photosynthetic plastids. According to the bulk flow hypothesis, genes are transferred to the nucleus following the disturbance of organelles. Disturbance was common in the early stages of endosymbiosis, however, once the host cell gained control of organelle division, eukaryotes could evolve to have only one plastid per cell. Having only one plastid severely limits gene transfer as the lysis of the single plastid would likely result in cell death. Consistent with this hypothesis, organisms with multiple plastids show an 80-fold increase in plastid to nucleus gene transfer compared to organisms with single plastids.

Evidence

Evidence that mitochondria and plastids arose from bacteria is as follows:

- New mitochondria and plastids are formed only through a process similar to binary fission. Binary fission is the form of cell division used by bacteria and archaea.
- If a cell's mitochondria or chloroplasts are removed, the cell does not have the means to create new ones. For example, in some algae, such as *Euglena*, the plastids can be destroyed by certain chemicals or prolonged absence of light without otherwise affecting the cell. In such a case, the plastids will not regenerate.
- Transport proteins called porins are found in the outer membranes of mitochondria and chloroplasts and are also found in bacterial cell membranes.
- A membrane lipid cardiolipin is exclusively found in the inner mitochondrial membrane and bacterial cell membranes.
- Some mitochondria and some plastids contain single circular DNA molecules that are similar to the DNA of bacteria both in size and structure.
- Genome comparisons suggest a close relationship between mitochondria and Rickettsial bacteria.
- Genome comparisons suggest a close relationship between plastids and cyanobacteria.
- Both mitochondria and plastids have small genomes compared to bacteria and neither organelle would be capable of surviving outside the cells where they are found. This is consistent with an increased dependence on the host after forming an endosymbiosis.
- Many genes in the genomes of mitochondria and chloroplasts have been lost or transferred to the nucleus of the host cell. Consequently, the chromosomes of many eukaryotes contain genes that originated from the genomes of mitochondria and plastids.
- Mitochondrial and plastid ribosomes are more similar those found in bacteria (70S) than those found in eukaryotes.
- Proteins created by mitochondria and chloroplasts use N-formylmethionine as the initiating amino acid, as do proteins created by bacteria but not proteins created by eukaryotic nuclear genes or archaea.
- Much of the internal structure and biochemistry of plastids (e.g. the presence of thylakoids and particular chlorophylls) is very similar to that of cyanobacteria.
- Similar plastids can be found in groups of algae that are not closely related to each other. This suggests that if chloroplasts originated *de novo*, they did so multiple times. If this were true, their close similarity to each other despite being present in distantly related algae is difficult to explain.

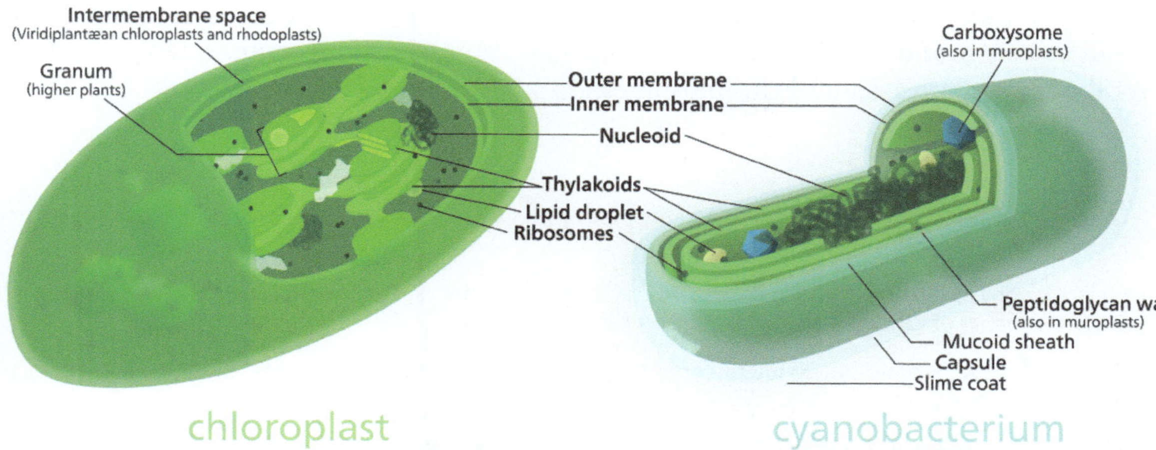

Comparison of chloroplasts and cyanobacteria showing their similarities.

Secondary Endosymbiosis

Primary endosymbiosis involves the engulfment of a bacterium by another free living organism. Secondary endosymbiosis occurs when the product of primary endosymbiosis is itself engulfed and retained by another free living eukaryote. Secondary endosymbiosis has occurred several times and has given rise to extremely diverse groups of algae and other eukaryotes. Some organisms can take opportunistic advantage of a similar process, where they engulf an alga and use the products of its photosynthesis, but once the prey item dies (or is lost) the host returns to a free living state. Obligate secondary endosymbionts become dependent on their organelles and are unable to survive in their absence. RedToL, the Red Algal Tree of Life Initiative funded by the National Science Foundation highlights the role red algae or Rhodophyta played in the evolution of our planet through secondary endosymbiosis.

One possible secondary endosymbiosis in process has been observed by Okamoto & Inouye (2005). The heterotrophic protist *Hatena* behaves like a predator until it ingests a green alga, which loses its flagella and cytoskeleton, while *Hatena*, now a host, switches to photosynthetic nutrition, gains the ability to move towards light and loses its feeding apparatus.

The process of secondary endosymbiosis left its evolutionary signature within the unique topography of plastid membranes. Secondary plastids are surrounded by three (in euglenophytes and some dinoflagellates) or four membranes (in haptophytes, heterokonts, cryptophytes, and chlorarachniophytes). The two additional membranes are thought to correspond to the plasma membrane of the engulfed alga and the phagosomal membrane of the host cell. The endosymbiotic acquisition of a eukaryote cell is represented in the cryptophytes; where the remnant nucleus of the red algal symbiont (the nucleomorph) is present between the two inner and two outer plastid membranes.

Despite the diversity of organisms containing plastids, the morphology, biochemistry, genomic organisation, and molecular phylogeny of plastid RNAs and proteins suggest a single origin of all extant plastids – although this theory is still debated.

Some species including *Pediculus humanus* have multiple chromosomes in the mitochondrion. This and the phylogenetics of the genes encoded within the mitochondrion suggest that mitochondria have multiple ancestors, that these were acquired by endosymbiosis on several occasions rather than just once, and that there have been extensive mergers and rearrangements of genes on the several original mitochondrial chromosomes.

References

- Lane, Nick (2005). Power, Sex, Suicide. Mitochondria and the Meaning of Life. New York: Oxford University Press. p. 14. ISBN 9780199205646.
- Corning, Peter A. (2010). Holistic Darwinism: Synergy, Cybernetics, and the Bioeconomics of Evolution. Chicago: University of Chicago Press. p. 81. ISBN 978-0-22611-633-4.
- Wise, Robert R; Hoober,, J. Kenneth (2007). Structure and function of plastids. Berlin: Springer. p. 104. ISBN 9781402065705. .
- Smith, A.E.; Marcker, K.A. (1968). "N-formylmethionyl transfer RNA in mitochondria from yeast and rat liver". J Mol Biol. 38 (2): 241. doi:10.1016/0022-2836(68)90409-9. Retrieved 18 May 2016.
- Okamoto, Noriko; Inouye, Isao (2005). "A Secondary Symbiosis in Progress?" (PDF). Science. 310 (5746): 287. doi:10.1126/science.1116125. PMID 16224014. Retrieved 15 February 2015.
- "Mitochondria Share an Ancestor With SAR11, a Globally Significant Marine Microbe". ScienceDaily. July 25, 2011. Retrieved 2011-07-26.
- J. Cameron Thrash; et al. (2011). "Phylogenomic evidence for a common ancestor of mitochondria and the SAR11 clade". Scientific Reports. 1: 13. doi:10.1038/srep00013. PMID 22355532.
- Raven, P., George Johnson, Kenneth Mason, Jonathan Losos, Susan Singer, 2010. Biology. McGraw-Hill 9th Ed. (January 14, 2010)
- Fischer K, Weber A, Brink S, Arbinger B, Schünemann D, Borchert S, Heldt HW, Popp B, Benz R, Link TA (1994). "Porins from plants. Molecular cloning and functional characterization of two new members of the porin family". J Biol Chem. 269 (41): 25754–25760. PMID 7523392.
- Zeth K, Thein M (2010). "Porins in prokaryotes and eukaryotes: common themes and variations". Biochem J. 431 (1): 13–22. doi:10.1042/BJ20100371. PMID 20836765.

8

Applied Fields of Cell Biology

Cell biology is very important in the fields of genetics and pharmaceutical sciences. Experiments on drug effectiveness and their side-effects as well as newer techniques of cloning and gene splicing all began with the study of cell biology. The topics discussed in the chapter are of great importance to broaden the existing knowledge on cell biology.

Cell Therapy

Cell therapy (also called cellular therapy or cytotherapy) is therapy in which cellular material is injected into a patient; this generally means intact, living cells. For example, T cells capable of fighting cancer cells via cell-mediated immunity may be injected in the course of immunotherapy.

Cell therapy originated in the nineteenth century when scientists experimented by injecting animal material in an attempt to prevent and treat illness. Although such attempts produced no positive benefit, further research found in the mid twentieth century that human cells could be used to help prevent the human body rejecting transplanted organs, leading in time to successful bone marrow transplantation.

Today two distinct categories of cell therapy are recognized.

The first category is cell therapy in mainstream medicine. This is the subject of intense research and the basis of potential therapeutic benefit. Such research can be controversial when it involves human embryonic material.

The second category is in alternative medicine, and perpetuates the practice of injecting animal materials in an attempt to cure disease. This practice, according to the American Cancer Society, is not backed by any medical evidence of effectiveness, and can have deadly consequences.

Background

Cell therapy can be defined as therapy in which cellular material is injected into a patient.

There are two branches of cell therapy: one is legitimate and established, whereby human cells are transplanted from a donor to a patient; the other is dangerous alternative medicine, whereby injected animal cells are used to attempt to treat illness.

The origins of cell therapy can perhaps be traced to the nineteenth century, when Charles-Édouard Brown-Séquard (1817–1894) injected animal testicle extracts in an attempt to stop the effects of aging. In 1931 Paul Niehans (1882–1971) – who has been called the inventor of cell therapy – attempted to cure a patient by injecting material from calf embryos. Niehans claimed to have treated many people for cancer using this technique, though his claims have never been validated by research.

Charles-Édouard Brown-Séquard – tried to stop aging by injecting animal testicle extract

In 1953 researchers found that laboratory animals could be helped not to reject organ transplants by pre-inoculating them with cells from donor animals; in 1968, in Minnesota, the first successful human bone marrow transplantation took place.

Bone marrow transplants have been found to be effective, along with some other kinds of human cell therapy – for example in treating damaged knee cartilage. In recent times, cell therapy using human material has been recognized as an important field in the treatment of human disease. The experimental field of Stem cell therapy has shown promise for new types of treatment.

In mainstream medicine, cell therapy is supported by a distinct healthcare industry which sees strong prospects for future growth.

In mainstream Medicine

Allogeneic Cell Therapy

In allogeneic cell therapy the donor is a different person to the recipient of the cells. In pharmaceutical manufacturing, the allogenic methodology is promising because unmatched allogenic therapies can form the basis of "off the shelf" products. There is research interest in attempting to develop such products to treat conditions including Crohn's disease and a variety of vascular conditions.

Human Embryonic Stem Cells

Research into human embryonic stem cells is controversial, and regulation varies from country to country, with some countries banning it outright. Nevertheless, these cells are being investigated as the basis for a number of therapeutic applications, including possible treatments for diabetes and Parkinson's disease.

Mechanisms of Action

Cell therapy is targeted at many clinical indications in multiple organs and by several modes of cell delivery. Accordingly, the specific mechanisms of action involved in the therapies are wide ranging. However, there are two main principles by which cells facilitate therapeutic action:

1. Stem cell or progenitor cell engraftment, differentiation, and long term replacement of damaged tissue. In this paradigm multipotent or unipotent cells differentiate into a specific cell type in the lab or after reaching the site of injury (via local or systemic administration). These cells then integrate into the site of injury, replacing damaged tissue, and thus facilitate improved function of the organ or tissue. An example of this is the use of cells to replace cardiomyocytes after myocardial infarction.

2. Cells that have the capacity to release soluble factors such as cytokines, chemokines, and growth factors which act in a paracrine or endocrine manner. These factors facilitate self-healing of the organ or region. The delivered cells (via local or systemic administration) remain viable for a relatively short period (days-weeks) and then die. This includes cells that naturally secrete the relevant therapeutic factors, or which undergo epigenetic changes or genetic engineering that causes the cells to release large quantities of a specific molecule. Examples of this include cells that secrete factors which facilitate angiogenesis, anti-inflammation, and anti-apoptosis. This mode of action is proposed by companies such as Pluristem and Pervasis that use adherent stromal cells or mature endothelial cells to treat peripheral artery disease and arteriovenous access complications.

Neural Stem Cell Therapy

Neural stem cells (NSCs) are the subject of ongoing research for possible therapeutic applications, for example for treating a number of neurological disorders such as Parkinson's disease and Huntington's disease.

Mesenchymal Stem Cell Therapy

MSCs are immunomodulatory, multipotent and fast proliferating and these unique capabilities mean they can be used for a wide range of treatments including immune-modulatory therapy, bone and cartilage regeneration, myocardium regeneration and the treatment of Hurler syndrome, a skeletal and neurological disorder.

Researchers have demonstrated the use of MSCs for the treatment of osteogenesis imperfecta (OI). Horwitz et al. transplanted bone marrow (BM) cells from human leukocyte antigen (HLA)-identical siblings to patients suffering from OI. Results show that MSCs can develop into normal osteoblasts, leading to fast bone development and reduced fracture frequencies. A more recent clinical trial showed that allogeneic fetal MSCs transplanted in utero in patients with severe OI can engraft and differentiate into bone in a human fetus.

Besides bone and cartilage regeneration, cardiomyocyte regeneration with autologous BM MSCs has also been reported recently. Introduction of BM MSCs following myocardial infarction (MI) resulted in significant reduction of damaged regions and improvement in heart function. Clinical trials for treatment of acute MI with Prochymal by Osiris Therapeutics are underway. Also, a clinical trial revealed huge improvements in nerve conduction velocities in Hurler's Syndrome patients infused with BM MSCs from HLA-identical siblings.

Hematopoietic Stem Cell Transplantation

HSCs possess the ability to self-renew and differentiate into all types of blood cells, especially those involved in the human immune system. Thus, they can be used to treat blood and immune disorders. Since human bone marrow (BM) grafting was first published in 1957, there have been significant advancements in HSCs therapy. Following that, syngeneic marrow infusion and allogeneic marrow grafting were performed successfully. HSCs therapy can also render its cure by reconstituting damaged blood-forming cells and restoring the immune system after high-dose chemotherapy to eliminate disease.

There are three types of HSCT: syngeneic, autologous, and allogeneic transplants. Syngeneic transplantations occur between identical twins. Autologous transplantations use the HSCs obtained directly from the patient and hence do not cause any complications of tissue incompatibility; whereas allogeneic transplantations involve the use of donor HSCs, either genetically related or unrelated to the recipient. To lower the risks of transplant, which include graft rejection and Graft-Versus-Host Disease (GVHD), allogeneic HSCT must satisfy compatibility at the HLA loci (i.e. genetic matching to reduce the immunogenicity of the transplant). Mismatch of HLA loci would result in treatment-related mortality and higher risk of acute GVHD.

In addition to BM derived HSCs, the use of alternative sources such as umbilical cord blood (UCB) and peripheral blood stem cells (PBSCs) has been increasing. In comparison with BM derived HSCs recipients, PBSCs recipients afflicted with myeloid malignancies reported a faster engraftment and better overall survival. However, this was at the expense of increased rate of GVHD. Also, the use of UCB requires less stringent HLA loci matching, although the time of engraftment is longer and graft failure rate is higher.

In Alternative Medicine

In alternative medicine, cell therapy is defined as the injection of non-human cellular animal material in an attempt to treat illness. Quackwatch labels this as "senseless", since "cells from the organs of one species cannot replace the cells from the organs of other species" and because a number of serious adverse effects have been reported.

Of this alternative, animal-based form of cell therapy, the American Cancer Society say: "Available scientific evidence does not support claims that cell therapy is effective in treating cancer or any other disease. In may in fact be lethal ...".

Cloning

In biology, **cloning** is the process of producing similar populations of genetically identical individuals that occurs in nature when organisms such as bacteria, insects or plants reproduce asexually. Cloning in biotechnology refers to processes used to create copies of DNA fragments (molecular cloning), cells (cell cloning), or organisms. The term also refers to the production of multiple copies of a product such as digital media or software.

Many organisms, including aspen trees, reproduce by cloning.

The term clone, invented by J. B. S. Haldane, is derived from the Ancient Greek word κλών *klōn*, "twig", referring to the process whereby a new plant can be created from a twig. In horticulture, the spelling *clon* was used until the twentieth century; the final *e* came into use to indicate the vowel is a "long o" instead of a "short o". Since the term entered the popular lexicon in a more general context, the spelling *clone* has been used exclusively.

In botany, the term lusus was traditionally used.

Natural Cloning

Cloning is a natural form of reproduction that has allowed life forms to spread for more than 50 thousand years. It is the reproduction method used by plants, fungi, and bacteria, and is also the way that clonal colonies reproduce themselves. Examples of these organisms include blueberry plants, hazel trees, the Pando trees, the Kentucky coffeetree, *Myricas*, and the American sweet-gum.

Molecular Cloning

Molecular cloning refers to the process of making multiple molecules. Cloning is commonly used to amplify DNA fragments containing whole genes, but it can also be used to amplify any DNA sequence such as promoters, non-coding sequences and randomly fragmented DNA. It is used in a wide array of biological experiments and practical applications ranging from genetic fingerprinting to large scale protein production. Occasionally, the term cloning is misleadingly used to refer to the identification of the chromosomal location of a gene associated with a particular phenotype of interest, such as in positional cloning. In practice, localization of the gene to a chromosome or genomic region does not necessarily enable one to isolate or amplify the relevant genomic sequence. To amplify any DNA sequence in a living organism, that sequence must be linked to an origin of replication, which is a sequence of DNA capable of directing the propagation of itself and any linked sequence. However, a number of other features are needed, and a variety of specialised cloning vectors (small piece of DNA into which a foreign DNA fragment can be inserted) exist that allow protein production, affinity tagging, single stranded RNA or DNA production and a host of other molecular biology tools.

Cloning of any DNA fragment essentially involves four steps

1. fragmentation - breaking apart a strand of DNA
2. ligation - gluing together pieces of DNA in a desired sequence
3. transfection - inserting the newly formed pieces of DNA into cells
4. screening/selection - selecting out the cells that were successfully transfected with the new DNA

Although these steps are invariable among cloning procedures a number of alternative routes can be selected; these are summarized as a *cloning strategy.*

Initially, the DNA of interest needs to be isolated to provide a DNA segment of suitable size. Subsequently, a ligation procedure is used where the amplified fragment is inserted into a vector (piece of DNA). The vector (which is frequently circular) is linearised using restriction enzymes, and incubated with the fragment of interest under appropriate conditions with an enzyme called DNA ligase. Following ligation the vector with the insert of interest is transfected into cells. A number of alternative techniques are available, such as chemical sensitivation of cells, electroporation, optical injection and biolistics. Finally, the transfected cells are cultured. As the aforementioned procedures are of particularly low efficiency, there is a need to identify the cells that have been successfully transfected with the vector construct containing the desired insertion sequence in the required orientation. Modern cloning vectors include selectable antibiotic resistance markers, which allow only cells in which the vector has been transfected, to grow. Additionally, the cloning vectors may contain colour selection markers, which provide blue/white screening (alpha-factor complementation) on X-gal medium. Nevertheless, these selection steps do not absolutely guarantee that the DNA insert is present in the cells obtained. Further investigation of the resulting colonies must be required to confirm that cloning was successful. This may be accomplished by means of PCR, restriction fragment analysis and/or DNA sequencing.

Cell Cloning

Cloning Unicellular Organisms

Cloning a cell means to derive a population of cells from a single cell. In the case of unicellular organisms such as bacteria and yeast, this process is remarkably simple and essentially only requires the inoculation of the appropriate medium. However, in the case of cell cultures from multi-cellular organisms, cell cloning is an arduous task as these cells will not readily grow in standard media.

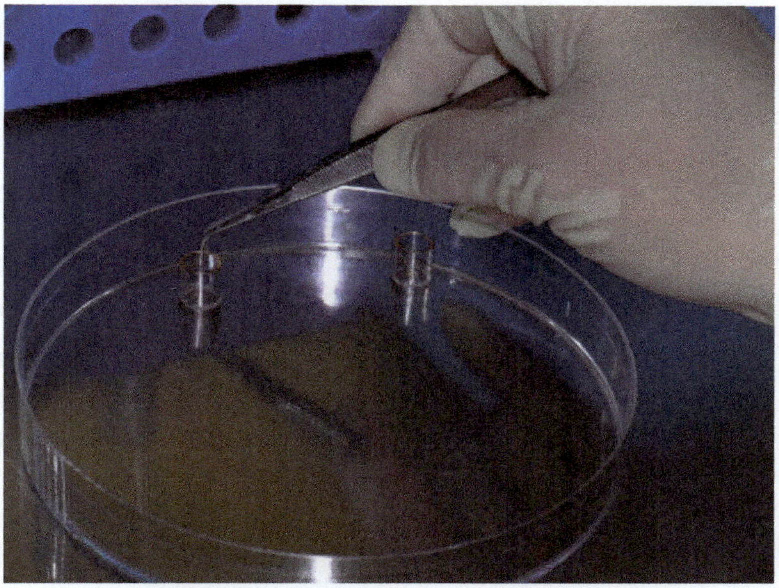

Cloning cell-line colonies using cloning rings

A useful tissue culture technique used to clone distinct lineages of cell lines involves the use of cloning rings (cylinders). In this technique a single-cell suspension of cells that have been exposed to a mutagenic agent or drug used to drive selection is plated at high dilution to create isolated colonies, each arising from a single and potentially clonal distinct cell. At an early growth stage when colonies consist of only a few cells, sterile polystyrene rings (cloning rings), which have been dipped in grease, are placed over an individual colony and a small amount of trypsin is added. Cloned cells are collected from inside the ring and transferred to a new vessel for further growth.

Cloning Stem Cells

Somatic-cell nuclear transfer, known as SCNT, can also be used to create embryos for research or therapeutic purposes. The most likely purpose for this is to produce embryos for use in stem cell research. This process is also called "research cloning" or "therapeutic cloning." The goal is not to create cloned human beings (called "reproductive cloning"), but rather to harvest stem cells that can be used to study human development and to potentially treat disease. While a clonal human blastocyst has been created, stem cell lines are yet to be isolated from a clonal source.

Therapeutic cloning is achieved by creating embryonic stem cells in the hopes of treating diseases such as diabetes and Alzheimer's. The process begins by removing the nucleus (containing the DNA) from an egg cell and inserting a nucleus from the adult cell to be cloned. In the case of someone with Alzheimer's disease, the nucleus from a skin cell of that patient is placed into an empty egg. The reprogrammed cell begins to develop into an embryo because the egg reacts with the transferred nucleus. The embryo will become genetically identical to the patient. The embryo will then form a blastocyst which has the potential to form/become any cell in the body.

The reason why SCNT is used for cloning is because somatic cells can be easily acquired and cultured in the lab. This process can either add or delete specific genomes of farm animals. A key point to remember is that cloning is achieved when the oocyte maintains its normal functions and instead of using sperm and egg genomes to replicate, the oocyte is inserted into the donor's somatic cell nucleus. The oocyte will react on the somatic cell nucleus, the same way it would on sperm cells.

The process of cloning a particular farm animal using SCNT is relatively the same for all animals. The first step is to collect the somatic cells from the animal that will be cloned. The somatic cells could be used immediately or stored in the laboratory for later use. The hardest part of SCNT is removing maternal DNA from an oocyte at metaphase II. Once this has been done, the somatic nucleus can be inserted into an egg cytoplasm. This creates a one-cell embryo. The grouped somatic cell and egg cytoplasm are then introduced to an electrical current. This energy will hopefully allow the cloned embryo to begin development. The successfully developed embryos are then placed in surrogate recipients, such as a cow or sheep in the case of farm animals.

SCNT is seen as a good method for producing agriculture animals for food consumption. It successfully cloned sheep, cattle, goats, and pigs. Another benefit is SCNT is seen as a solution to clone endangered species that are on the verge of going extinct. However, stresses placed on both the egg cell and the introduced nucleus can be enormous, which led to a high loss in resulting cells in early research. For example, the cloned sheep Dolly was born after 277 eggs were used for SCNT, which created 29 viable embryos. Only three of these embryos survived until birth, and only one survived to adulthood. As the procedure could not be automated, and had to be performed manually under a microscope, SCNT was very resource intensive. The biochemistry involved in reprogramming the differentiated somatic cell nucleus and activating the recipient egg was also far from being well-understood. However, by 2014 researchers were reporting cloning success rates of seven to eight out of ten and in 2016, a Korean Company Sooam Biotech was reported to be producing 500 cloned embryos per day.

In SCNT, not all of the donor cell's genetic information is transferred, as the donor cell's mitochondria that contain their own mitochondrial DNA are left behind. The resulting hybrid cells retain those mitochondrial structures which originally belonged to the egg. As a consequence, clones such as Dolly that are born from SCNT are not perfect copies of the donor of the nucleus.

Organism Cloning

Organism cloning (also called reproductive cloning) refers to the procedure of creating a new multicellular organism, genetically identical to another. In essence this form of cloning is an asexual method of reproduction, where fertilization or inter-gamete contact does not take place. Asexual reproduction is a naturally occurring phenomenon in many species, including most plants and some insects. Scientists have made some major achievements with cloning, including the asexual reproduction of sheep and cows. There is a lot of ethical debate over whether or not cloning should be used. However, cloning, or asexual propagation, has been common practice in the horticultural world for hundreds of years.

Horticultural

The term *clone* is used in horticulture to refer to descendants of a single plant which were produced by vegetative reproduction or apomixis. Many horticultural plant cultivars are clones, having been derived from a single individual, multiplied by some process other than sexual reproduction. As an example, some European cultivars of grapes represent clones that have been propagated for over two millennia. Other examples are potato and banana. Grafting can be regarded as cloning, since all the shoots and branches coming from the graft are genetically a clone of a single individual, but this particular kind of cloning has not come under ethical scrutiny and is generally treated as an entirely different kind of operation.

Many trees, shrubs, vines, ferns and other herbaceous perennials form clonal colonies naturally. Parts of an individual plant may become detached by fragmentation and grow on to become separate clonal individuals. A common example is in the vegetative reproduction of moss and liverwort gametophyte clones by means of gemmae. Some vascular plants e.g. dandelion and certain viviparous grasses also form seeds asexually, termed apomixis, resulting in clonal populations of genetically identical individuals.

Parthenogenesis

Clonal derivation exists in nature in some animal species and is referred to as parthenogenesis (reproduction of an organism by itself without a mate). This is an asexual form of reproduction that is only found in females of some insects, crustaceans, nematodes, fish (for example the hammerhead shark), the Komodo dragon and lizards. The growth and development occurs without fertilization by a male. In plants, parthenogenesis means the development of an embryo from an unfertilized egg cell, and is a component process of apomixis. In species that use the XY sex-determination system, the offspring will always be female. An example is the little fire ant (*Wasmannia auropunctata*), which is native to Central and South America but has spread throughout many tropical environments.

Artificial Cloning of Organisms

Artificial cloning of organisms may also be called *reproductive cloning*.

First Moves

Hans Spemann, a German embryologist was awarded a Nobel Prize in Physiology or Medicine in 1935 for his discovery of the effect now known as embryonic induction, exercised by various parts of the embryo, that directs the development of groups of cells into particular tissues and organs. In 1928 he and his student, Hilde Mangold, were the first to perform somatic-cell nuclear transfer using amphibian embryos – one of the first moves towards cloning.

Methods

Reproductive cloning generally uses “somatic cell nuclear transfer” (SCNT) to create animals that are genetically identical. This process entails the transfer of a nucleus from a donor adult cell (somatic cell) to an egg from which the nucleus has been removed, or to a cell from a blastocyst from which the nucleus has been removed. If the egg begins to divide normally it is transferred into the uterus of the surrogate mother. Such clones are not strictly identical since the somatic cells may contain mutations in their nuclear DNA. Additionally, the mitochondria in the cytoplasm also contains DNA and during SCNT this mitochondrial DNA is wholly from the cytoplasmic donor’s egg, thus the mitochondrial genome is not the same as that of the nucleus donor cell from which it was produced. This may have important implications for cross-species nuclear transfer in which nuclear-mitochondrial incompatibilities may lead to death.

Artificial *embryo splitting* or *embryo twinning*, a technique that creates monozygotic twins from a single embryo, is not considered in the same fashion as other methods of cloning. During that procedure, an donor embryo is split in two distinct embryos, that can then be transferred via embryo transfer. It is optimally performed at the 6- to 8-cell stage, where it can be used as an expansion of IVF to increase the number of available embryos. If both embryos are successful, it gives rise to monozygotic (identical) twins.

Dolly the Sheep

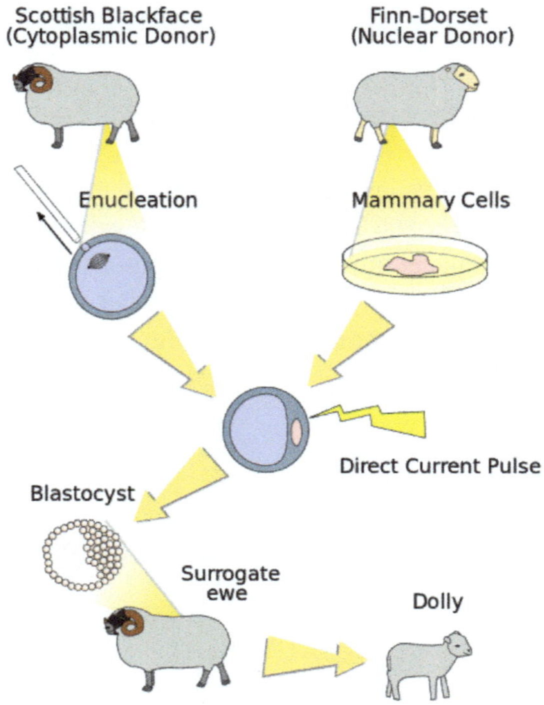

Dolly clone

Dolly, a Finn-Dorset ewe, was the first mammal to have been successfully cloned from an adult somatic cell. Dolly was formed by taking a cell from the udder of her 6-year old biological mother. Dolly's embryo was created by taking the cell and inserting it into a sheep ovum. It took 434 attempts before an embryo was successful. The embryo was then placed inside a female sheep that went through a normal pregnancy. She was cloned at the Roslin Institute in Scotland and lived there from her birth in 1996 until her death in 2003 when she was six. She was born on 5 July 1996 but not announced to the world until 22 February 1997. Her stuffed remains were placed at Edinburgh's Royal Museum, part of the National Museums of Scotland.

Dolly was publicly significant because the effort showed that genetic material from a specific adult cell, programmed to express only a distinct subset of its genes, can be reprogrammed to grow an entirely new organism. Before this demonstration, it had been shown by John Gurdon that nuclei from differentiated cells could give rise to an entire organism after transplantation into an enucleated egg. However, this concept was not yet demonstrated in a mammalian system.

The first mammalian cloning (resulting in Dolly the sheep) had a success rate of 29 embryos per 277 fertilized eggs, which produced three lambs at birth, one of which lived. In a bovine experiment involving 70 cloned calves, one-third of the calves died young. The first successfully cloned horse, Prometea, took 814 attempts. Notably, although the first clones were frogs, no adult cloned frog has yet been produced from a somatic adult nucleus donor cell.

There were early claims that Dolly the sheep had pathologies resembling accelerated aging. Scientists speculated that Dolly's death in 2003 was related to the shortening of telomeres, DNA-protein complexes that protect the end of linear chromosomes. However, other researchers, including Ian Wilmut who led the team that successfully cloned Dolly, argue that Dolly's early death due to respiratory infection was unrelated to deficiencies with the cloning process. This idea that the nuclei have not irreversibly aged was shown in 2013 to be true for mice.

Dolly was named after performer Dolly Parton because the cells cloned to make her were from a mammary gland cell, and Parton is known for her ample cleavage.

Species Cloned

The modern cloning techniques involving nuclear transfer have been successfully performed on several species. Notable experiments include:

- Tadpole: (1952) Robert Briggs and Thomas J. King had successfully cloned northern leopard frogs: thirty-five complete embryos and twenty-seven tadpoles from one-hundred and four successful nuclear transfers.
- Carp: (1963) In China, embryologist Tong Dizhou produced the world's first cloned fish by inserting the DNA from a cell of a male carp into an egg from a female carp. He published the findings in a Chinese science journal.
- Mice: (1986) A mouse was successfully cloned from an early embryonic cell. Soviet scientists Chaylakhyan, Veprencev, Sviridova, and Nikitin had the mouse "Masha" cloned. Research was published in the magazine "Biofizika" volume XXXII, issue 5 of 1987.
- Sheep: Marked the first mammal being cloned (1984) from early embryonic cells by Steen Willadsen. Megan and Morag cloned from differentiated embryonic cells in June 1995 and Dolly the sheep from a somatic cell in 1996.
- Rhesus monkey: Tetra (January 2000) from embryo splitting
- Pig: the first cloned pigs (March 2000). By 2014, BGI in China was producing 500 cloned pigs a year to test new medicines.
- Gaur: (2001) was the first endangered species cloned.
- Cattle: Alpha and Beta (males, 2001) and (2005) Brazil
- Cat: CopyCat "CC" (female, late 2001), Little Nicky, 2004, was the first cat cloned for commercial reasons
- Rat: Ralph, the first cloned rat (2003)
- Mule: Idaho Gem, a john mule born 4 May 2003, was the first horse-family clone.

- Horse: Prometea, a Haflinger female born 28 May 2003, was the first horse clone.
- Dog: Snuppy, a male Afghan hound was the first cloned dog (2005).
- Wolf: Snuwolf and Snuwolffy, the first two cloned female wolves (2005).
- Water buffalo: Samrupa was the first cloned water buffalo. It was born on 6 February 2009, at India's Karnal National Diary Research Institute but died five days later due to lung infection.
- Pyrenean ibex (2009) was the first extinct animal to be cloned back to life; the clone lived for seven minutes before dying of lung defects.
- Camel: (2009) Injaz, is the first cloned camel.
- Pashmina goat: (2012) Noori, is the first cloned pashmina goat. Scientists at the faculty of veterinary sciences and animal husbandry of Sher-e-Kashmir University of Agricultural Sciences and Technology of Kashmir successfully cloned the first Pashmina goat (Noori) using the advanced reproductive techniques under the leadership of Riaz Ahmad Shah.
- Gastric brooding frog: (2013) The gastric brooding frog, *Rheobatrachus silus*, thought to have been extinct since 1983 was cloned in Australia, although the embryos died after a few days.

Human Cloning

Human cloning is the creation of a genetically identical copy of a human. The term is generally used to refer to artificial human cloning, which is the reproduction of human cells and tissues. It does not refer to the natural conception and delivery of identical twins. The possibility of human cloning has raised controversies. These ethical concerns have prompted several nations to pass legislature regarding human cloning and its legality.

Two commonly discussed types of theoretical human cloning are *therapeutic cloning* and *reproductive cloning*. Therapeutic cloning would involve cloning cells from a human for use in medicine and transplants, and is an active area of research, but is not in medical practice anywhere in the world, as of 2014. Two common methods of therapeutic cloning that are being researched are somatic-cell nuclear transfer and, more recently, pluripotent stem cell induction. Reproductive cloning would involve making an entire cloned human, instead of just specific cells or tissues.

Ethical Issues of Cloning

There are a variety of ethical positions regarding the possibilities of cloning, especially human cloning. While many of these views are religious in origin, the questions raised by cloning are faced by secular perspectives as well. Perspectives on human cloning are theoretical, as human therapeutic and reproductive cloning are not commercially used; animals are currently cloned in laboratories and in livestock production.

Advocates support development of therapeutic cloning in order to generate tissues and whole organs to treat patients who otherwise cannot obtain transplants, to avoid the need for immunosuppressive drugs, and to stave off the effects of aging. Advocates for reproductive cloning believe that parents who cannot otherwise procreate should have access to the technology.

Opponents of cloning have concerns that technology is not yet developed enough to be safe and that it could be prone to abuse (leading to the generation of humans from whom organs and tissues would be harvested), as well as concerns about how cloned individuals could integrate with families and with society at large.

Religious groups are divided, with some opposing the technology as usurping "God's place" and, to the extent embryos are used, destroying a human life; others support therapeutic cloning's potential life-saving benefits.

Cloning of animals is opposed by animal-groups due to the number of cloned animals that suffer from malformations before they die, and while food from cloned animals has been approved by the US FDA, its use is opposed by groups concerned about food safety.

Cloning Extinct and Endangered Species

Cloning, or more precisely, the reconstruction of functional DNA from extinct species has, for decades, been a dream. Possible implications of this were dramatized in the 1984 novel *Carnosaur* and the 1990 novel *Jurassic Park*. The best current cloning techniques have an average success rate of 9.4 percent (and as high as 25 percent) when working with familiar species such as mice,[note 1] while cloning wild animals is usually less than 1 percent successful. Several tissue banks have come into existence, including the "Frozen Zoo" at the San Diego Zoo, to store frozen tissue from the world's rarest and most endangered species.

In 2001, a cow named Bessie gave birth to a cloned Asian gaur, an endangered species, but the calf died after two days. In 2003, a banteng was successfully cloned, followed by three African wildcats from a thawed frozen embryo. These successes provided hope that similar techniques (using surrogate mothers of another species) might be used to clone extinct species. Anticipating this possibility, tissue samples from the last *bucardo* (Pyrenean ibex) were frozen in liquid nitrogen immediately after it died in 2000. Researchers are also considering cloning endangered species such as the giant panda and cheetah.

In 2002, geneticists at the Australian Museum announced that they had replicated DNA of the thylacine (Tasmanian tiger), at the time extinct for about 65 years, using polymerase chain reaction. However, on 15 February 2005 the museum announced that it was stopping the project after tests showed the specimens' DNA had been too badly degraded by the (ethanol) preservative. On 15 May 2005 it was announced that the thylacine project would be revived, with new participation from researchers in New South Wales and Victoria.

In January 2009, for the first time, an extinct animal, the Pyrenean ibex mentioned above was cloned, at the Centre of Food Technology and Research of Aragon, using the preserved frozen cell nucleus of the skin samples from 2001 and domestic goat egg-cells. The ibex died shortly after birth due to physical defects in its lungs.

One of the most anticipated targets for cloning was once the woolly mammoth, but attempts to extract DNA from frozen mammoths have been unsuccessful, though a joint Russo-Japanese team is currently working toward this goal. In January 2011, it was reported by Yomiuri Shimbun that a team of scientists headed by Akira Iritani of Kyoto University had built upon research by Dr. Wakayama, saying that they will extract DNA from a mammoth carcass that had been preserved in a Russian laboratory and insert it into the egg cells of an African elephant in hopes of producing a mammoth embryo. The researchers said they hoped to produce a baby mammoth within six years. It was noted, however that the result, if possible, would be an elephant-mammoth hybrid rather than a true mammoth. Another problem is the survival of the reconstructed mammoth: ruminants rely on a symbiosis with specific microbiota in their stomachs for digestion.

Scientists at the University of Newcastle and University of New South Wales announced in March 2013 that the very recently extinct gastric-brooding frog would be the subject of a cloning attempt to resurrect the species.

Many such "de-extinction" projects are described in the Long Now Foundation's Revive and Restore Project.

Lifespan

After an eight-year project involving the use of a pioneering cloning technique, Japanese researchers created 25 generations of healthy cloned mice with normal lifespans, demonstrating that clones are not intrinsically shorter-lived than naturally born animals.

In a detailed study released in 2016 and less detailed studies by others suggest that once cloned animals get past the first month or two of life they are generally healthy. However, early pregnancy loss and neonatal losses are still greater with cloning than natural conception or assisted reproduction (IVF). Current research endeavors are attempting to overcome this problem.

In Popular Culture

In an article in the 8 November 1993 article of *Time*, cloning was portrayed in a negative way, modifying Michelangelo's *Creation of Adam* to depict Adam with five identical hands. *Newsweek*'s 10 March 1997 issue also critiqued the ethics of human cloning, and included a graphic depicting identical babies in beakers.

Cloning is a recurring theme in a wide variety of contemporary science fiction, ranging from action films such as *Jurassic Park* (1993), *The 6th Day* (2000), *Resident Evil* (2002), *Star Wars* (2002) and *The Island* (2005), to comedies such as Woody Allen's 1973 film *Sleeper*.

Science fiction has used cloning, most commonly and specifically human cloning, due to the fact that it brings up controversial questions of identity. *A Number* is a 2002 play by English playwright Caryl Churchill which addresses the subject of human cloning and identity, especially nature and nurture. The story, set in the near future, is structured around the conflict between a father (Salter) and his sons (Bernard 1, Bernard 2, and Michael Black) – two of whom are clones of the first one. *A Number* was adapted by Caryl Churchill for television, in a co-production between the BBC and HBO Films.

A recurring sub-theme of cloning fiction is the use of clones as a supply of organs for transplantation. The 2005 Kazuo Ishiguro novel *Never Let Me Go* and the 2010 film adaption are set in an alternate history in which cloned humans are created for the sole purpose of providing organ donations to naturally born humans, despite the fact that they are fully sentient and self-aware. The 2005 film *The Island* revolves around a similar plot, with the exception that the clones are unaware of the reason for their existence.

The use of human cloning for military purposes has also been explored in several works. *Star Wars* portrays human cloning in *Clone Wars*.

The exploitation of human clones for dangerous and undesirable work was examined in the 2009 British science fiction film *Moon*. In the futuristic novel *Cloud Atlas* and subsequent film, one of the story lines focuses on a genetically-engineered fabricant clone named Sonmi~451 who is one of millions raised in an artificial "wombtank," destined to serve from birth. She is one of thousands of clones created for manual and emotional labor; Sonmi herself works as a server in a restaurant. She later discovers that the sole source of food for clones, called 'Soap', is manufactured from the clones themselves.

Cloning has been used in fiction as a way of recreating historical figures. In the 1976 Ira Levin novel *The Boys from Brazil* and its 1978 film adaptation, Josef Mengele uses cloning to create copies of Adolf Hitler.

In 2012, a Japanese television show named "Bunshin" was created. The story's main character, Mariko, is a woman studying child welfare in Hokkaido. She grew up always doubtful about the love from her mother, who looked nothing like her and who died nine years before. One day, she finds some of her mother's belongings at a relative's house, and heads to Tokyo to seek out the truth behind her birth. She later discovered that she was a clone.

In the 2013 television show *Orphan Black*, cloning is used as a scientific study on the behavioral adaptation of the clones. In a similar vein, the book *The Double* by Nobel Prize winner José Saramago explores the emotional experience of a man who discovers that he is a clone.

References

- *Rantala, Milgram, M., Arthur (1999). Cloning: For and Against. Chicago, Illinois: Carus Publishing Company. p. 1. ISBN 0-8126-9375-2.*
- *Peter J. Russel (2005). iGenetics: A Molecular Approach. San Francisco, California, United States of America: Pearson Education. ISBN 0-8053-4665-1.*
- *Zastrow, Mark (8 February 2016). "Inside the cloning factory that creates 500 new animals a day". New Scientist. Retrieved 23 February 2016.*
- Shukman, David (14 January 2014) China cloning on an 'industrial scale' BBC News Science and Environment, Retrieved 27 February 2016
- *Baer, Drake (8 September 2015). "This Korean lab has nearly perfected dog cloning, and that's just the start". Tech Insider, Innovation. Retrieved 27 February 2016.*
- Shukman, David (14 January 2014) China cloning on an 'industrial scale' BBC News Science and Environment, Retrieved 10 April 2014
- Sagers, Larry A. (2 March 2009) Proliferate favorite trees by grafting, cloning Utah Stae University, Deseret News (Salt Lake City) Retrieved 21 February 2014
- Shubin, Neil (24 February 2008) Birds Do It. Bees Do It. Dragons Don't Need To New York Times, Retrieved 21 February 2014

- Shukman, David (14 January 2014) China cloning on an 'industrial scale' BBC News Science and Environment, Retrieved 14 January 2014
- *"Generations of Cloned Mice With Normal Lifespans Created: 25th Generation and Counting". Science Daily. 7 March 2013. Retrieved 8 March 2013.*
- *"Yvonne A. De La Cruz "Science Fiction Storytelling and Identity: Seeing the Human Through Android Eyes"" (PDF). Retrieved 2012-08-19.*
- Oh SKW and Choo ABH (2011) The Biological Basis | Stem Cells. In: Murray Moo-Young (ed.), Comprehensive Biotechnology, Second Edition, Volume 1, pp. 341–365. Elsevier.
- *Halim, N. (September 2002). "Extensive new study shows abnormalities in cloned animals". Massachusetts institute of technology. Retrieved October 2011.* Check date values in: |access-date= (help)
- *"Cloning Fact Sheet". Human Genome Project Information. Archived from the original on 2 May 2013. Retrieved 25 October 2011.*
- *Havstad JC (2010). "Human reproductive cloning: a conflict of liberties.". Bioethics. 24 (2): 71–7. doi:10.1111/j.1467-8519.2008.00692.x. PMID 19076121.*

Permissions

All chapters in this book are published with permission under the Creative Commons Attribution Share Alike License or equivalent. Every chapter published in this book has been scrutinized by our experts. Their significance has been extensively debated. The topics covered herein carry significant information for a comprehensive understanding. They may even be implemented as practical applications or may be referred to as a beginning point for further studies.

We would like to thank the editorial team for lending their expertise to make the book truly unique. They have played a crucial role in the development of this book. Without their invaluable contributions this book wouldn't have been possible. They have made vital efforts to compile up to date information on the varied aspects of this subject to make this book a valuable addition to the collection of many professionals and students.

This book was conceptualized with the vision of imparting up-to-date and integrated information in this field. To ensure the same, a matchless editorial board was set up. Every individual on the board went through rigorous rounds of assessment to prove their worth. After which they invested a large part of their time researching and compiling the most relevant data for our readers.

The editorial board has been involved in producing this book since its inception. They have spent rigorous hours researching and exploring the diverse topics which have resulted in the successful publishing of this book. They have passed on their knowledge of decades through this book. To expedite this challenging task, the publisher supported the team at every step. A small team of assistant editors was also appointed to further simplify the editing procedure and attain best results for the readers.

Apart from the editorial board, the designing team has also invested a significant amount of their time in understanding the subject and creating the most relevant covers. They scrutinized every image to scout for the most suitable representation of the subject and create an appropriate cover for the book.

The publishing team has been an ardent support to the editorial, designing and production team. Their endless efforts to recruit the best for this project, has resulted in the accomplishment of this book. They are a veteran in the field of academics and their pool of knowledge is as vast as their experience in printing. Their expertise and guidance has proved useful at every step. Their uncompromising quality standards have made this book an exceptional effort. Their encouragement from time to time has been an inspiration for everyone.

The publisher and the editorial board hope that this book will prove to be a valuable piece of knowledge for students, practitioners and scholars across the globe.

Index